the **timberframe** way

BOOKS BY MICHAEL MORRIS
AND DICK PIROZZOLO
Timberframe Plan Book

ALSO BY DICK PIROZZOLO
AND LINDA CORZINE
Timberframe Interiors

the **timberframe** way

A lavishly illustrated guide to the most elegant way to build a home

MICHAEL MORRIS AND **DICK PIROZZOLO**

The Lyons Press

Guilford, Connecticut

An imprint of The Globe Pequot Press

The Lyons Press is an imprint of The Globe Pequot Press

10 9 8 7 6 5 4 3 2 1

Printed in Great Britain
Designed by LeAnna Weller Smith

Library of Congress Cataloging-in-Publication Data

Pirozzolo, Dick, 1944-
  The timberframe way : a lavishly illustrated guide to the most elegant
way to build a home / Dick Pirozzolo and Michael Morris.
      p. cm.
Includes index.
  ISBN 1-59228-150-8 (HC : alk. paper)
  1. Wooden-frame houses. 2. Log buildings. 3. Timber. 4.
Architecture, Domestic--United States. I. Morris, Michael, 1947- II.
Title.
  TH4818.W6.P577 2003
  690'.837--dc22
                        2003015018

contents

ACKNOWLEDGMENTS:

The authors gratefully acknowledge Jay Cassell, editorial director of The Lyons Press. They especially thank Lisa Purcell, their editor, for her hard work, patience, and insightful editing and designer LeAnna Weller Smith for making this a beautiful book.

the **timberframe** way

angels in the architecture

CLASSIC POST & BEAM/
JIM BATTLES

PREVIOUS PAGE LEFT:
DAVIS FRAME COMPANY/
RICH FRUTCHEY

PREVIOUS PAGE RIGHT:
CARGILL/BLAKE ARCHITECTS &
BUILDERS, CLARK PLANNING &
DESIGN/SANDY AGRAFIOTIS

why timberframes?

What separates a timberframe home from other types of house structures? More to the point, what makes them special—unique among residential architectural styles and construction methods?

Timberframe buildings are actually simpler than most structures, with larger frame members that result in fewer individual parts. This oversized, often massive, framing does the work of all the wall studs, roof rafters, and innumerable other, relatively smaller pieces of dimensional lumber used in standardized "stick-frame" homebuilding. The entire structural skeleton of a typical Timberframe may have fewer than two hundred individual timbers. This simplicity is part of its appeal—by reducing the construction to its most basic elements, you end up with a greater whole. "Less is more," as architect Mies van der Rohe observed.

Today, Timberframes represent a unique segment among all of the residences built each year across the country, and their appeal is widespread and steadily growing. Many new homes borrow elements of timber framing to enhance their livability and distinctiveness. What timberframe homes offer is an alternative to conventional structures, as well as a home-focused architectural glimpse of history, grandeur, and even our own social and cultural identity.

HEARTHSTONE, INC.
*While the low roofline of a Colonial
saltbox-inspired design offers a
modest curbside appearance, the
multi-level interior is spacious and
designed to take advantage of
spectacular views.*

the evolution
of timber framing
in america

Timber framing was an intrinsic part of our nation's historic begin-
nings. In manufacturing terms it was a step above log construction,
and as such was a favored building method of the more established
or prosperous settlers of the colonial era.

Most home styles from this time are studies in evolution. Few early
settlers could afford to build what they wanted, when they wanted
it. People made do in the beginning with smaller, more basic
dwellings, and then extended outward or added architectural
details as their fortunes and families grew. For example, the classic
center-hall Colonial–house style, as we know it today, has a fireplace
at each end and a main staircase in the middle (in the central hall).
But in an era when everything from lumber to hardware to window
glass was handmade and in short supply, often only half of this
structure was built initially. The house at first had its entry hall and
stairs located at the front left or right corner, much like the common
row houses of that day. At a later date, an additional wing, along
with a second fireplace, completed the symmetrical plan.

Saltbox-style houses are another popular early New England architectural form that "grew" with the times. Saltboxes, distinguished by
their peaked gables at each end and a long, sloping roofline in back,
developed from a basic upright rectangle shape—the signature
sloped roof came later, the result of a convenient shed extension

added after the original house was built. In some cases, this extension started out as a rude shed that was open to the elements and used as a shelter for animals, then eventually was enclosed to serve as a storage room, and still later—as the family grew—as additional living space.

Anyone who visits a home preserved intact from our colonial era will see that timber framing invariably was used in its construction. These early American residences are living history books that tell the story of how our nation was settled by farmers, merchants, and workers from abroad. Depending on when and where the homes were built, they are either a pure reflection of the settlers' own heritage or a synthesis of influences from many countries.

According to George Senerchia, a Timberframe "collector" who restores and researches early examples of the craft, the seventeenth-century English who settled in the Northeast were typical of the immigrants who brought their traditional building techniques to the New World. Senerchia knows his subject intimately, having rebuilt American timber structures from every century since the Europeans first established their colonies on our shores. Like an archaeologist, he has examined the "bones" of timberframe buildings dating back 400 years or more.

The oldest structure this hands-on historian has worked on dates from 1680, and it was built based on the principles that influenced these settlers. Few of them were landowners in their native countries, and wood structures in that time and place had immense value because only the landed aristocracy had access to the forests and the sturdy hardwoods that grew in them. Wood was a precious resource not to be squandered, and the lesson was not wasted on the settlers after they arrived in the New World. This can be seen in early New England examples of English timberframe construction where a single post was used at an intersection of several beams. To create a post that could carry the weight and width of these beams, which were often secured to one another by complex joinery without nails or other fasteners, the builders literally turned a tree

upside down—putting the widest end of the post at the top. This
"English tying joint" is still used today, and the vertical timber that
supports it is descriptively known as a "jowled" or "gunstock" post.

Later, Dutch timber-building techniques were introduced in
Pennsylvania and upstate New York, and these structures also
reflected the sensibilities of their designers. These farmers and
dairymen needed taller structures that allowed for dry hay and
feed storage above the ground floor where the animals were kept.
Erecting a full-height second story required significantly greater
investments of engineering, materials, and labor, but the ever-
practical Dutch instead developed an alternative. In their barns,
the first-floor wall posts were cut longer to extend a few feet
above the ceiling joists, creating a loft storage area while providing
just enough headroom for workmen to move around easily. These
knee-high walls, or "knee-walls," are still commonly used to reduce
building costs while increasing available space in bedroom attics
as well as haylofts.

Examples such as these of timberframe construction and adapta-
tion are visible throughout the history of architecture in America.
In fact, structural visibility is key to this unique, elemental style of
construction. Unlike most other building methods, the framing in
a timber home is, by design, left completely exposed—the timbers are on view for occupants and visitors to enjoy. The beauty
and grandeur of this network of massive, often hand-hewn timbers inside your living room or kitchen, or throughout your home,
cannot be understated. And it's not just for historic or period-style
homes. The most architecturally modern homes today are often
either timberframe structures or incorporate elements of timber
framing in their design.

**EUCLID TIMBER FRAMES/
JOHN PHIPPEN**

*In contrast to the soaring spaces
of the great room, this cozy nook
borrows elements of Victoriana
punctuated with natural leather
upholstery.*

**EUCLID TIMBER FRAMES/
JOHN PHIPPEN**

*This spacious great room relies on
massive beams and natural stone
for impact.*

FROM TUDOR
TO YOUR DOOR

Just because you live here doesn't
mean you have to limit your
home-building ambitions to
American architectural designs.
A British company, Border Oak
Design & Construction Limited,
imports timberframe homes to
the United States that are true to
their Old English architectural
roots, and completely fashioned
from English oak, one of the
most enduring and traditional
of woods.

Border Oak will cut and ship an
exact replica of Shakespeare's
cottage at Stratford-on-Avon, if
that's your midsummer-night's
dream home, or a full-blown
English country manse, for those
who prefer a bit more ostentation.
These homes are crafted to
exacting detail right up to their
rooftops. They are even available
with genuine thatched roofs.
The roofs—laboriously crafted of
layered, interwoven natural reeds
or thatch—were common on
these "uncommon" structures
when the original homes were
built. For buyers who insist on
absolute authenticity, the company
will even send over a professional
thatcher to complete the
installation.

To its credit, one of Border Oak's
goals is to revive the forgotten
skills of Tudor craftsmen and draw
inspiration from the sixteenth

century, when Elizabethan
England enjoyed enormous
prosperity and reveled in new
ideas of style and comfort. The
company founders skillfully
combine the quaint construction
methods of that time with a
selective use of modern technology,
which includes gaskets to ensure
an airtight and watertight fit
where beams join walls.

Established in 1980 in rural
Herefordshire, the heartland of a
locally prominent building style
known as "half-timber" framing,
Border Oak produces about one
hundred structures a year. The
builders still use the Scribe Rule
marking system, and every frame
is completely assembled at their
facility after it is cut, then disas-
sembled and stacked for delivery.

BORDER OAK DESIGN & CONSTRUCTION LIMITED

Depending on your taste and budget, Border Oak Design offers Americans the opportunity to own a genuine English Tudor Home.

methods of construction

TIMBERFRAME VS. POST AND BEAM

The terms "timberframe" and "post and beam" are interchangeable and often confused. A Timberframe is, in essence, a structure built mainly of solid-wood posts and beams that are typically left exposed in the interior. Purists, however, insist that true timberframe construction also employs traditional methods and materials fastened and fitted together in the time-honored way—with interlocking, often hand-hewn joinery that eschews modern mechanical fasteners of any sort. Timberframe joinery, in fact, is typically secured only with wooden pegs called *trunnels* (an ancient contraction of the words *tree* and *nail*).

Because this massive framework is on view for all to see, carpenters over the centuries have elevated their craft with a blend of elegance and rusticity. Jeff Davis, of the Davis Frame Company in Claremont, New Hampshire, notes that fabricators routinely add special touches to "dress" their wood, using techniques such as "skip-planing," which imparts a random, handmade look to beam surfaces, or complex intricacies such as "gunstock" posts and joinery that rivals fine furniture construction. Sometimes this finish-work has a practical side as well. Posts and beams are often chamfered or routed along their edges, for example. This gives them an elegant, finished appearance, but it also makes the edges less sharp where occupants may come in contact with them, as in doorways or in stairwells.

Alternately, post-and-beam construction is far more straightforward, often employing square-cut dimensional lumber and even

**THISTLEWOOD TIMBER
FRAME HOMES**

In addition to mixing various wood species, this owner cut a graceful arc in the corner braces, which is a low-cost way to add curves to an otherwise linear structure.

JACK SOBON

Together, peeled twig balusters, minimally hewn curved beams, and a massive central fireplace make this home look as though early English settlers had built it.

"engineered" lumber such as laminated beams and trusses. This building style also allows for metal fasteners such as steel bolts or plates and more simplified wood joinery. In some cases, wood plugs or pegs hide the fasteners, or the plates are incorporated into the truss designs to appear ornamental, making the final effect similar to timberframe construction.

Some post-and-beam manufacturers, such as Classic Post & Beam in York, Maine, offer buyers the option of including timberframe-style joinery where it matters, such as in an elaborate hammerbeam truss that is a major focal point in the home. Less-costly metal fasteners are then used in other areas of the home where the joinery is not as prominent.

PRE-CUT, SYSTEMS-BUILT, AND DELIVERED TO YOUR DOOR

Timberframe homes are usually custom designed and constructed, and in general cost more to build than standard houses because of the limited availability of both heavy timbers and the builders experienced in creating them. Fabricating a complex framework of massive timbers on-site would be a daunting task for most contractors. As a result, few Timberframes are actually built this way. Most are prefabricated, ready-to-assemble building systems.

Rather than ship whole uncut, unjoined timbers to a building site, the majority of today's timber framers opt for frames pre-cut in facilities equipped to easily handle and precisely fashion them according to a home's blueprints. Modern machinery can joint cross-sectional pieces up to twelve or more inches thick and equally wide, making even the most difficult compound cuts to tolerances within fractions of an inch, so it's no wonder that factory-made frames are often considered better built than those hewn "in the field."

That's not to say that all of the cutting and fitting is complete when a frame arrives at your site. To allow for variations in site conditions, such as the relative straightness of a foundation's walls, posts are often cut extra long and shipped with "theoretical bottoms" that can then be trimmed to fit. Frame sections that lock together using trunnels or splines also sometimes require a bit of hands-on job-site work to mate successfully, often due to wood shrinkage, swelling, or twisting after they are cut.

LEFT:

**EUCLID TIMBER FRAMES/
JOHN PHIPPEN**

*Stone and natural wood
harmonize perfectly in this
Rocky Mountain manor.*

Still, to avoid surprises at the building site, most fabricators test-fit their frame sections after cutting. Typically, the bents and various joints are assembled flat on a spacious shop floor, and then the sections are dismantled and each timber is "branded" with a chalk mark or incised symbol that corresponds to its place in the building schedule. When the parts are shipped to the building site, they are frequently loaded onto trucks last-piece-first, so that the timbers needed in the earliest stages of construction are at the top of the load and can be put to use as soon as the shipment arrives.

This method of prefabricating frame parts has helped to reduce the high cost of "custom" Timberframes in a number of ways. By standardizing and pre-cutting certain pieces used in most frames, such as roof purlins, corner braces, and even whole truss sections, manufacturers can offer catalogs of less-expensive "off-the-shelf" designs, or incorporate these standard cuts into custom designs wherever possible to shave dollars off the bottom line.

STRUCTURAL INSULATED PANELS

During the timber framing revival of the 1960s and 1970s, many frames were closed in with exterior walls of traditional two-by-four stick-frame construction—which proved both costly and problematic for two reasons. First, if the frames were wrapped in conventional framing, this amounted to building an extra house around the structural timbers. Second, stud walls built between the timbers often eventually resulted in gaps when the timbers and dimensional materials naturally swelled, shrank, or settled at different rates.

Coincidentally, around that time a type of sandwich-like exterior "stressed skin" panel was evolving. These panels incorporated

ABOVE:
THE MURUS COMPANY

Rigid foam filled structural insulating panels (SIPs) used to enclose modern timberframe homes are fastened together using a camlock system. The convenient wiring chase accommodates electrical, phone, cable, and even fiber-optic service.

CLASSIC POST & BEAM

This Colonial offers all the benefits of post-and-beam construction on the interior, while the exterior is designed to harmonize perfectly in a neighborhood setting of quality custom homes. The narrow clapboards enhance the traditional New England style.

"SQUARE RULE" AND "SCRIBE RULE"

Back in the early days of timber framing, builders cut and assembled their frames as they went along, using their own idiosyncratic methods to keep track of each beam or bent. Each builder had a personal marking system—often kept secret from others who might steal his valuable knowledge of this highly specialized trade. The problem with this "system" was that it slowed production and limited the amount of work the builder and only his most trusted helpers could accomplish. Just think of the difficulty if a framer died halfway through a big project— assembling his frame would be a nightmare for anyone not schooled in his methods.

Later, when framing skills became common, builders saw the value in creating a system that everyone could learn and use no matter where a frame was raised. First they created the Square Rule, in which all pieces were cut square on their *exterior* faces. This created a dependable reference point for all building measurements as well as for the placement and alignment of outer walls. Additionally, angles for specific frame members were standardized—ninety degrees for corners and forty-five degrees for braces.

**NORTHFORD TIMBERFRAMERS/
DICK PIROZZOLO**

*Race knives—timber framers used
these indispensable scribing tools
to mark structural elements so
that they could be kept organized
when moved from the shop to the
building site.*

The builders further refined this
system by adding what was called
the Scribe Rule. Using a device
called a "race knife," which was
part caliper and part scribing tool,
they marked the timbers wherever
they intersected. As the builder
test-assembled the parts, he
would stick the caliper point
in one of the timbers and inscribe
an arc across the joint onto both
pieces. Next, the timbers were
each marked with matching
Roman numerals. When the time
came for final assembly, all the
workers had to do was match up
the numbers and the markings,
and the frame pieces went together
as they were originally intended.

*The repetition of bents creates depth
and charm, leading the eye through
this hallway. The stone floor and
antiqued timbers are reminiscent
of old Spanish missions.*

WHEN BUYING A TIMBERFRAME HOME

Some manufacturers provide a
package of components required
to complete the basic structure,
including dimensional framing for
interior walls and floors, plywood
subflooring, exterior walls, and a
roof system. Others offer "frame-
only" packages that include little
more than the timbers. Each has
its benefits and drawbacks in
terms of cost and complexity for
the buyer.

Many timber framers construct,
or *raise*, the frame and leave the
completion of the house up to
the consumer. If you choose this
method, before you sign a
contract you should arrange a
meeting or other type of direct
communication between the
manufacturer and your builder,
carpenters, and relevant subcontractors such as the foundation
mason and electrician. This is the
time to make sure that all parties
will work well together and that
your workers understand the
requirements and methodology
of timberframe construction.

plywood or oriented-strand board (OSB) sheathing bonded to solid
foam insulation, creating one-piece wall sections that exceeded the
strength and insulating value of comparable size two-by-four or
two-by-six walls. Now generically known as "structural insulated
panels," or SIPs, they are the preferred method for enclosing timberframe homes today.

In many ways, SIPs are marvels of convenience. A home's entire
exterior, including walls and roof, can be pre-cut at the factory and
shipped to its site for speedy erection. Door and window openings
are often pre-cut by computer-aided machinery to guarantee
precision. Interior finish materials such as drywall and paneling can
also be specified and installed on SIPs prior to delivery. To simplify
construction, most panels are designed with integral channels, or
chases, to accommodate electrical, cable, and telephone wires. The
panels are so strong that they can be used alone to create house
sections without timber-framing support, and their rigid foam cores
generally provide higher insulation values than competing materials
such as fiberglass or cellulose.

According to one manufacturer, The Murus Company, "SIPs also
act as a moisture barrier and virtually eliminate energy-robbing
air infiltration, drafts, and uneven room temperatures." The foam
used in most SIPs (typically polyisocyanurate or expanded poly-
styrene) is free of formaldehyde or other harmful compounds such
as CFCs and is safe to humans and the environment.

Because these panels can be self-supporting, SIPs are often used to
bring the cost of a Timberframe within reach. In a so-called "hybrid"
structure, the most visible areas of the home (such as master bedrooms, kitchens, and "public" areas such as living rooms and entry halls)

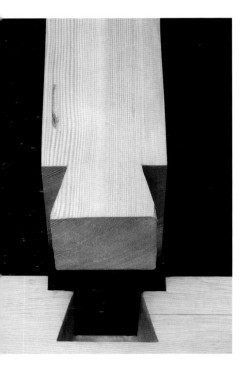

are designed with a handsome timberframe interior, while other less prominent sections (such as bathrooms, small bedrooms, and utility areas) are built with less expensive SIPs. Would-be buyers who want the timberframe home of their dreams yet can't afford a true custom structure should discuss such options with their builder and timberframe or post-and-beam manufacturer.

timberframe terminology

Without question, building or owning a true Timberframe is even more fun when you can speak in the vernacular when discussing this ancient art and craft. Most of the terminology evolved long ago from early Gothic and Anglo-Saxon builders, and the words and descriptions we use today are still redolent with meaning. The word *trunnel*, for example, combines *tree* and *nail* to perfectly explain the hardwood pegs used as timber-framing fasteners—sort of a land-lubber's equivalent to gunnel (gun wale) or boson (boat swain).

Here are some of the most common terms used to describe the principal parts of a timberframe home:

post. The "legs" of every frame and the principal weight-bearing frame members. The earliest posts were simply trees cut to length and stood on end. Anything vertical in a frame is often referred to as a post.

beam. Any horizontal timber. Specialized beams have their own nomenclature: A *summerbeam* is a main structural member used to carry the weight of floor joists or to tie girts together; a *sill* is a foundation beam typically used around the bottom, or perimeter, of the frame.

girt. The principal beam (from which we get the word *girder*) that spans the length and sometimes the width of a house.

bent. Timberframes are typically organized and built in sections, with each section divided by a flat, wall-like arrangement of posts, beams, and rafters called a bent. Bents extend completely through the structure, from sidewall to sidewall and roof peak to floor, similar to bread slices in a loaf. Each endwall is a bent, and between each pair of bents is a bay, or area of the house. In construction, bent sections are assembled flat on the floor deck, tipped up or raised in place with the help of a crane, then connected one to another to form the house.

king post. The vertical center post in a truss, used to unify and support the truss beams, chords, and other frame members. It is typically suspended from the roof ridge or other upper structure and rarely reaches the floor.

queen post. Used when a truss is configured with two symmetrical posts, with or without a central king post.

truss. A strong geometric framing assembly designed to span open areas and transfer the weight of a roof or other structure, such as an upper floor or balcony, to the walls. Trusses come in a wide variety of shapes and styles depending on the load they carry, the complexity of the structure, and the art of the builder. Notable examples include the X-shaped "scissor" truss and the gracefully arched "hammerbeam" truss.

purlin. A horizontal beam typically used to interconnect and strengthen roof rafters, or to serve as an intermediate framing member for fastening other materials such as interior walls or paneling.

beadle. When early builders needed something big enough to coax or beat (bead) stubborn timbers into place, they reached for the biggest hammer they could find. This tool was—and still is—often fashioned from a timber cutoff or even a length of natural log, and is wielded by the brawniest men in the crew.

CHAPTER TWO

turning "house" into home

TIMBERPEG/RICH FRUTCHEY

A broad roofline, generous eves, natural materials, and innovative window placement are among the design strategies used to create a low profile for a home that blends perfectly with its environment.

PREVIOUS PAGE LEFT: LINDAL CEDAR HOMES, INC.

PREVIOUS PAGE RIGHT: CLASSIC POST & BEAM/ SANDY AGRAFIOTIS

speaking in the vernacular

The word "vernacular" is most often used to describe how or what people commonly speak—the local *lingua franca*, the language of a particular region or group. The term is frequently applied to architecture as well, usually to mean, "what people around here like to build." Although local architecture, like everything else in America, has been plagiarized and homogenized almost beyond recognition of its primal roots, there was a time when you could actually tell where you stood by looking at the type of homes and buildings around you. New England had its compact, vertical Saltboxes and Capes. In the Southeast along the hurricane coast, generous roofs, storm shutters, and salt-weathered siding typified low-country houses. Farther south and southwest, stucco walls and tile roofs evoked a Spanish colonial past. Meanwhile, true Westerners had their horizontally expansive ranch houses to remind themselves, and others, of their cowboy heritage. And up in the northern states, Norwegian settlers—called "tie hackers" because they cut ties for the railroads out of local forests—introduced their own brand of timber-building styles strongly influenced by the abundant forests and sense of limitless resources America offered.

According to Jack Sobon, a Berkshire, Massachusetts, resident, registered architect, and author of *Historic American Timber Joinery: A Graphic Guide*, "Vernacular homes are modest, demonstrate an appreciation of local materials and craftsmanship, and feel comfortable with the environment—teepees and igloos are, strictly speaking, vernacular homes. Timber framing is an extension of that love of

HEARTHSTONE, INC.

*The chinked log wall completes
the Western motif of this master
bedroom. Note the upper portion
is plaster.*

**BORDER OAK DESIGN &
CONSTRUCTION LIMITED**

*The dramatic truss work in these
homes recalls the cathedral ceilings
of bygone days. Post-and-beam
construction defines the space for
the homeowner's art collection.*

craft, local materials and sense
of place [resulting in] a home
that appears to be friendly,
inviting, and at peace with its
environment."

Sobon explains that, "Old builders sited and oriented their houses to take the best advantage of natural systems. Sun, prevailing winds, groundwater, soils, vegetation, and drainage were all considered since great amounts of energy are necessary to alter or compensate for site deficiencies. Because buildings had to work with their environment, they were naturally comfortable with their site, and in so doing made the occupants more comfortable inside."

WHEN "NATURAL" MEANT BASIC

Commenting on the back-to-nature
revival of the 1970s, the late Edward
C. Allen, who grew up on a family
farm in Westfield, Massachusetts,
recalled when the "natural" life
was anything but peaceful and
relaxed:

"Every night as I went to bed my
father would call up and say
'Edward! First thing in the morning
I want you to fill the woodbox!'
All night long I had that woodbox
in my mind, thinking, *I've got to
fill the woodbox, I've got to fill
the woodbox, I've got to fill the
woodbox!* You see, back then, living
was a full-time occupation."

a natural cycle

The idea that "natural" processes hold intrinsic value is something
that, over time, goes in and out of fashion in every society. Many
social and economic factors contributed to the rebirth of timber
framing, but one of the most significant was the "back-to-nature"
movement that began at the fringes of American society during
the 1960s and took hold among the middle class during the 1970s.
This was at a time when industry was migrating farther from core
cities to the suburbs, and beyond to exurbia. New outer beltways
constructed to accommodate this ever-expanding industrial development now bypassed the original ring highways, or beltways, built
around our burgeoning cities.

And who was moving out to exurbia? The growing ranks of young
urban professionals who were then on the cutting edge of the
computer age. Instead of adopting the trappings of slick modernism,
they quietly rebelled—seeking contrast from their work life by
turning inward to home, gardening, do-it-yourself-crafts, cooking
with organic foods, wearing natural fibers, and otherwise "getting
back to nature." All with help from PBS television that made home
economics programming a family event—the seventies' equivalent
of *The Ed Sullivan Show*.

This return to craftsmanship and self-reliance reflected our redis-
covered appreciation of hands-on values and natural, "home-grown"
materials. Timberframes and, to some degree, log homes spoke in
this vernacular and became ideal choices for the "new pioneers" of
the Woodstock generation.

AN OLD IDEA MADE NEW

This return to the vernacular of comfortable familiarity, of natural materials and handcrafted works, seems to occur whenever anything alien and new—such as the technologies of the modern age—threatens to overpower us. In the 1970s, it was the advent of the computer and its insidious encroachment on daily life. In 1919, the mechanized mayhem of World War I inspired the German Bauhaus movement, which de-emphasized "sentimental" architectural embellishment in favor of cold, impersonal industrial materials. This resulted in a quintessentially American reaction home-style led by Frank Lloyd Wright, one of the most visionary architects of the twentieth century.

PERIOD STYLING VS. ARCHITECTURAL ODDITIES

Mike MacPhail, a soft-spoken Belmont, Massachusetts, architect with Canadian roots and an easygoing manner, rankles nonetheless when it comes to poorly executed—or deliberately bad—home design.

"One thing that bothers me about modern homes is that their façades are so flat. Older homes—even styles that were purposely designed to have a flat front—had some dimensionality to the surface. Their windows were set in, they had sills that jutted out from the façade, and there were corner details, generous moldings, eaves, and cornices. You don't see that today, and the result is façades that lack texture and interest."

Not all architects and builders are well-versed in historic styles, MacPhail allows. They borrow from, but don't really follow, classic designs. "Although architecture does evolve, and most homes are a synthesis of various styles and eras, too many of today's homes, even expensive homes, are a clash of styles—Georgian and Victorian and Colonial and Federalist all thrown together in a mishmash."

Home Interiors also require architectural discipline, he adds, and good design is especially important in timberframe homes, where

Wright championed the idea that all man-made structures—homes, public offices, and commercial buildings alike—should reflect and incorporate elements of their surroundings, and be very much a part of their living environment. Wright's influence extended throughout architecture and is still with us today, and it can be seen in many modern timberframe designs. His "prairie-style" homes, created as showcases for *Ladies' Home Journal* in 1901, are classic examples. Although they emphasized flat rather than peaked roofs—the better to harmonize with the broad, end-less horizons of the windswept Plains—these homes were noted for their exposed structural elements, bold design lines, large overhanging eaves, massive central chimneys, and wide-open interior spaces.

An even earlier example of "reactionary" architecture can be seen in the nineteenth-century Arts and Crafts movement led by William Morris in England. In 1877, Morris heralded the importance of the craftsman and the inherent beauty of his handiwork in his lecture "The Decorative Arts and Their Relation to Modern Life and Progress." His philosophy was simple: "Men have so long delighted in…forms and intricacies that do not necessarily imitate Nature," he said, "but in which the hand of the craftsman is guided to work in the way that She does."

It is no coincidence that this back-to-nature movement occurred during the onset of the Industrial Age—an era that saw stunning advances in technology coupled with equally stunning horrors spawned by dehumanizing mass-production, as machinery replaced the hand of man and God in life's work. Arts and Crafts–style homes, later exemplified by the architectural team of Greene and Greene in America, again elevated handiwork to its rightful place, this time

open-plan layouts magnify design details and design flaws. "I don't like to see the end of a wall that is only as wide as a two-by-four with two layers of wallboard. Walls need to terminate with mass to (visually) stabilize them in enormous volumes of space. When you end a wall in the middle of an open area, build in a bookshelf or some other detail. I would also beware of 'gotta-have-it' fads, such as Palladian windows. Even the great Italian architect Andrea Palladio didn't use them as often as we do today."

MacPhail reminds clients that any period-style home can be built using a timberframe structure, but he urges them to choose an architecture style that works well with this building method. "The structure should allow visual continuity between the interior and exterior parts of the building. In other words, the exterior elements should give the visitor a clue to the craftsmanship that is inside."

with a remarkable enthusiasm for over-the-top decorative design elements across a spectrum of organic materials that included interior fabrics, metals, stone, and, of course, wood.

So it is also no wonder that the art and craft of timber framing returned as a popular home-construction method, as well as an architectural style in its own right, in the twentieth century. In an era of high technology and high anxiety, the rustic, organic comfort of a well-timbered dwelling brings us a sense of peace and protection where we need it most.

LINDAL CEDAR HOMES, INC.
This well-designed home, with its abundance of windows offers natural light to the interior, while its open balconies and enclosed porch makes for comfortable outdoor living.

a plan that suits you

Familiarizing oneself with the various architectural styles can help when it comes to designing a home that will be pleasing to look at and live in. But it's also worth remembering that, except for the need to meet structural and building codes, there are no hard and fast design rules in residential construction.

How, then, can you be sure that the home you build will be comfortable to live in, with inviting interior spaces, good traffic patterns, cozy seating, and pleasant visual aspects from room to room? One way is to start with a "known" plan layout, a design your builder or architect has used before, a tried-and-true blueprint that has met the criteria for livability and passed with flying colors. In many cases, you can visit a model of the home you intend to build, walk through, and decide for yourself which elements meet your needs and match your lifestyle, and which you would modify.

But what if you are designing from scratch—working from a clean sheet of paper, so to speak? What assurances do you have that your dream house will be the comfortable, homey, sunlight-filled place you always hoped it would be—and not some gloomy, cramped, uninviting dungeon with a floor plan you ultimately find unlivable?

For boatwrights, there is a critical moment when, after months of design and construction, the hull is ready for its first (and sometimes final) test: Will it float? If the designer and builder have done their jobs properly, there is little doubt that it will. But a more important question for all concerned is: How *well* will it float? Will it skip over

DAVIS FRAME COMPANY/
RICH FRUTCHEY

*A single-story home has much
more character and an open
airy feeling when timberframe
construction is used.*

LET'S HEAR IT FOR THE RANCH

When planning a timberframe home, do not reject the ranch. Many people look down on this elemental building design because it is so, well…simple. Yet, the same attributes these design mavens dismiss are significant in what they offer.

The basic rectangular ranch is one of the easiest and least-complicated structures to build, which can help keep construction costs down. Its single-story layout may seem one-dimensional at first glance, but for elderly or disabled residents this is often a major advantage. A low-slung structure also may be the best option in a setting that calls for uninterrupted views, or on a site that requires a low profile.

In the hands of an inventive architect, ranches offer innumerable possibilities for design details. Gables and dormers may be added to accentuate the roofline or bring additional light and visual interest to interiors. Giving the roof ample pitch can turn a ranch into a chateau-style home with loft space and dramatic vaulted ceilings.

the waves and dance like a ballerina in light airs—or be a leaden scow that plows through the swell with a tendency to induce seasickness in its unhappy passengers?

This is the intangible unknown that all designers contend with. In many ways, every unique, custom-designed building is like that boat. And the world of architecture is filled with examples of beautiful buildings that, despite their designers' best intentions, turned out to be monuments to poor planning and lack of foresight.

Although there may be no firm rules, there are guidelines that have developed over the years. Anyone contemplating building their own house should seek out this information and at least familiarize themselves with the basics of good homebuilding. Of course, there are whole libraries filled with books on the subject, but one of the most interesting and thought provoking is *A Pattern Language*, written in 1977 by renowned architect Christopher Alexander and his colleagues at the Center for Environmental Structure at the University of California, Berkeley. This book attempts to identify the many tangible, as well as intangible, aspects of well-designed dwellings to help builders understand what makes houses—and professional buildings and even towns and cities—more or less livable. Why are some homes cozy and inviting, while others are dismal and off-putting? Believe it or not, the answer to this and other abstract questions of this sort can be found here.

While your next home may be located anywhere in the country, another book that provides universal strategies for building and maintaining a house in harmony with its environment is *Building with Nantucket in Mind*. Authors Christopher Lang and Kathy Stout are highly regarded authorities on nineteenth-century American

residential buildings. On Nantucket, where the entire island is designated as a Historic District, *The Book* (as it is reverentially known among builders and local officials) weighs in with valuable advice on everything from choosing appropriate shingles to the proper size and placement of windowpanes to designing streetscapes.

Whatever path you choose to attaining your timberframe home, above all remember that building one's own house is as much a journey as a destination. Make it a path that you enjoy, fill it with personal memories, and celebrate the craft of the tradesmen and artisans you encounter in the process. You'll get much more out of it, and you'll enjoy living in your home all the more.

Wider eaves, "chunky" architectural-grade or handsplit cedar shingles, and built-up exterior fascias accentuate this rusticated look. Even minor alterations to a ranch's basic floor plan, such as moving the main entry from its typical central location to an end wall, can alter the way a home "lives" and functions in a significant way.

GOING TO THE CHAPEL: MIKE AND KAREN'S STORY

People with only a passing familiarity with timberframe construction may remark that it offers too little variation in style, or even that "most" timberframe interiors look alike. Perhaps this is because, for many people, the only real points of reference they have are the occasional timbered structures they have seen or spent any time in, such as rustic resorts, historic barns, and churches. It's easy to understand why they have this reaction. Monumental wooden structures usually incorporate timber framing because big buildings require big structural supports. Churches—especially the larger and invariably older cathedrals—wouldn't have been possible without it. In fact, some of the best living examples of this building style today are ancient religious or historic buildings in Europe and Japan.

The first timberframe home my wife, Karen, and I built nearly ended up with a very churchlike interior. This happened even though, as a carpenter and stick-frame homebuilder, I thought I knew a thing or two about construction design. But my first attempt at designing a frame didn't go smoothly.

We had researched several frame manufacturers and decided to go with Riverbend Timber Framing of Blissfield, Michigan. At that time, in 1990, the company's reputation was already well established, which gave me confidence for my first "solo" undertaking. Their frames were solid oak, which both Karen and I liked. And the staff at Riverbend, from company president, Frank Baker, on down, were wonderfully patient and helpful, something we greatly appreciated.

Riverbend was eager to please us, so the conversation went something like this:

RB: What do you want your Timberframe to look like?

Us: Well, we really don't know. We know how big we want the house to be, and we have a pretty good idea of how we want it to lay out—where the bedrooms will go, how the kitchen should face east, the fact that we want a great room with a separate dining area, how we intend to site the house on the property—all of that. But because we've never designed or built a Timberframe, we have no clue as to what "our" framework should actually look like.

RB: Okay, then. We'll take your basic layout and design a framework into it. You can review it and tell us what you like and what you don't like, and we'll adjust from there.

Us: Sounds like a plan. Go to it.

So, after a week or two we received Riverbend's plans. Although they had done a creditable job designing a Timberframe complete with mighty posts and beams, high peaks, and soaring arches, our first reaction wasn't very positive. Those heavy posts came down into the middle of our cherished, wide-open living areas. The arches made the family room look like the nave of a church. There was too much wood in the ceilings for my wife's tastes, and so on.

We called Riverbend, somewhat apologetically, and tried to explain our feelings.

RB: No problem. Now that we know what you don't like about this plan, we can start working toward something that suits you better.

And that was really the starting point for us. Riverbend was great, showing us a number of iterations that steered us, step by step, ever closer to the timberframe home of our dreams. The posts were moved out of the way, replaced by trusses that could span our twenty-six-foot-wide great room without support. The arches became more angular and no longer looked as though they belonged in a church. Drywall was substituted for the cedar planking between the roof purlins, which made Karen happy and also made the overhead areas

seem higher, more open, and less imposing than they would have been in all wood.

Riverbend even followed my suggestion to increase the length and strength of the spans by using scissors trusses, and then went me one better by adding king posts—each one an incredibly complex feat of carpentry that forms the intersection of eleven separate timbers. Those king posts have become the "signature" pieces in our framework, never failing to elicit comments and compliments from visitors to our home.

It should be pointed out here that the word "king post," like so many other terms we use today to define the various parts and processes of timber framing, harks back to the medieval period when timber trusses and arches defined those soaring, magnificent spaces in the Gothic cathedrals of the day. So in a roundabout way it's not difficult to understand the connection some people see between churches and all other timberframe structures. Perhaps it's inevitable, even complimentary, that this connection exists, given the fact that some of the most beautiful, inspiring buildings throughout history still resonate so strongly within this construction style.

**THISTLEWOOD TIMBER
FRAME HOMES**

*Timberframe structural themes
may be echoed in the landscaping,
as this owner did by incorporating
a graceful arched wood footbridge
into the garden setting.*

INSIDE-OUTSIDE DESIGN SENSE

ON THE EXTERIOR:

• Architects and designers often draw rear elevations as an
afterthought, but the back of a house should be as pleasing as
the front, especially if the plans call for a rear deck or walk-out
basement leading to a patio, yard, or garden. After all, families
and their guests typically spend most of their quality time out
there.

• Patios and under-deck areas also become more attractive
spaces when they get equal design attention. And it's usually
easier and less costly to include improvements such as patio
pavers, stone walls, and privacy screening in the original construction budget than to add them later.

• Small design touches can make a big difference. Build up or
box tall deck columns, or use heavier posts, to avoid that
spindly look. Make sure windows and doors are properly positioned and "balanced" in your exterior plan. Moving a window
or door a few inches may improve the view from outside
without compromising the interior.

TIMBERPEG/RICH FRUTCHEY

Lofts are great spaces to get away to work on hobbies and crafts while staying tuned into the tempo of family of activities.

THISTLEWOOD TIMBER FRAME HOMES

Easels, stunning custom floors and cabinets, towering timbers, and ample space and light truly make this loft a "room with a view."

**EUCLID TIMBER FRAMES/
JOHN PHIPPEN**

*Art needn't be displayed formally—
it can instead become a moveable
feast for the eyes, as this homeowner
has done using shelves and even
the floor to accommodate an
extensive collection.*

YOU'RE NOT GETTING YOUNGER

If you're thinking of building a timberframe retirement or second home, Dana Delano of Ward Log Homes has some timely advice. She organized a group of senior citizens and home design experts to talk about what they wanted most in their dream home:

- A single-story floor plan eliminates the need to climb stairs. If design or site considerations make a second story necessary, a spacious master bedroom on the main floor is preferred, with guest bedrooms upstairs.

- An easy flow from interior rooms to decks, porches, and patios encourages a casual retirement lifestyle, as well as indoor-outdoor entertaining.

- Allow plenty of wall space and even a "trophy" room for displaying a lifetime of collections, family photos, and artwork.

- Convenience features such as entry ramps, lever door handles, and other ADA (Americans with Disabilities Act) requirements should be seamlessly blended into the interior and exterior design and the landscaping.

- Include a place for guests and grandchildren to sleep over, but don't make this area so spacious or so comfortable that they will overstay their welcome!

- If you know that your home's everyday entrance will be a side door, make it more inviting and perhaps spend less time and attention—and budget—on a grand front entryway that will seldom be used.

IN THE INTERIOR:

- Wide-open, high-volume interiors are easily created with post-and-beam structures, and lofts are literally the highpoints of timberframe living. Tucked away among the beams, lofts make for intimate, cozy spaces if you plan them that way. They're perfect for occasional use as guest rooms, library and reading areas, or hobby centers. Rooms with high ceilings also offer unique opportunities. For example, a sleeping loft built above a bath or a closet can double the usable area of a small bedroom.

- Not every room in a Timberframe lends itself to high ceilings. Lowering the overhead better accommodates task lighting in task-oriented spaces such as kitchens, and some owners simply prefer a more intimate setting in bedrooms and other private areas.

- Designers create intimacy by using darker colors or materials. Wood paneling between overhead beams will visually "lower" a ceiling, while white or neutral color drywall will make the same space appear higher and more open.

**CLASSIC POST & BEAM/
SANDY AGRAFIOTIS**

*An eclectic mix of furniture injects
the owner's personality into this
Craftsman-style home and adds an
air of informality to the décor of
this post-and-beam living room.*

*Dormers, gables, and roofline brows
are echoed in the distant mountain
peaks that dominate the view.*

VACATION PLANNING:
DICK AND JANE'S SPOT

When we built our timberframe
summer home—a dream that took
a dozen years with plenty of false
starts—we made our share of mis-
takes, but we did some things
right, too. Everyone knows, for
example, that you should try to
avoid designing a home while
you're building it, but sometimes
a little flexibility pays off.

Well into construction we
rethought the utility room space
that was supposed to accommodate our clothes washer and
dryer, hot water tank, and off-sea-
son storage. We said to ourselves,
"This is too nice—let's make it a
bedroom."

So our builder figured out a way to
squeeze the hot water tank under
the stairs, and Jane found a stacked
washer/dryer compact enough to
move in with it. It was a great deci-
sion. The extra bedroom is something we couldn't live without now.
If only it were a foot larger!

We did get carried away with
some things, though. In our zeal
for authenticity (our vacation
home is in a historic area), we
opted for a cedar-shingled roof.
It's nice, but not as practical as
the basement we could have had
if we had chosen less expensive
asphalt roofing.

(continued on page 52)

• Because natural wood is already a design focus in timberframe
homes, you can create greater visual impact by contrasting it
with other textures and materials, such as colorful art accents,
upholstered furniture, glass block, metals in a variety of finishes, and painted or glazed kitchen cabinets.

at home in
your environment

Positioning, or siting, your home on your property can be as important as designing the home itself. In our first book, *Timberframe
Home Plans*, we offered a great deal of advice on this part of the
process. After all, unlike painting a home a different color or even
relocating a window, choosing where the house sits is not something you can easily change after the fact.

If you have the time before you build, get out and walk the land.
Bring along your survey or plot plan, a 100-foot steel rule, some
string or brightly colored builders tape, a bagful of ground stakes,
and a mallet. It's best to view wooded property in both winter and
summer. Choosing a building site in summer when the trees are in
full leaf deprives you of valuable insights into how the surroundings
will affect the house when the trees are bare and the sun traces a
lower path in the sky.

Consider all the options for your site, not only where the house will
sit but also which direction it will face. Because the first floor will be
as much as four feet above grade, bring a ladder to check out the

view from the ground-floor window height. For a view from the second story, climb a tree or build a scaffold for a preview.

Once you have a pretty good idea of its location, stake out the perimeter of the house and outline it with string or tape to approximate its footprint. This will help to reveal any flaws in its relative position, as well as that of the driveway, decks, gardens, and so forth. Take note of the trees nearby, and mark those you'll want to keep for shading and those you will need to remove for safety. Trees standing near your building site can show you where the shadows of the house will fall, which will have an impact on the positioning of deck or garden areas.

Also bring along a compass with 360-degree markings on its dial. This can help you avoid misunderstandings with the builder later on. Saying "I want a southern exposure," isn't as accurate as telling him you want the kitchen windows to face 187 degrees south. Even a few degrees one way or the other can make a big difference in how a house is positioned on its site—not to mention how happy, or unhappy, you will be looking out your kitchen window for years to come.

When choosing a site, keep in mind that sometimes what seems like the best spot for your home can be just the opposite. For example, building atop the highest elevation may offer stunning vistas, but positioning the house just below the crest will move it out of the prevailing winds and shelter it into the terrain—while preserving, rather than interrupting, the view.

THISTLEWOOD
TIMBER FRAME HOMES

Stately evergreens frame this mountain retreat making it a part of the forest.

Here are few other lessons about vacation homes we learned along the way:

• If your home is built over a crawl space, insulating the floor will make it more comfortable and energy efficient year-round. Protect the insulation underneath the floor with plywood or wire mesh to discourage wildlife. In winter, surround an open perimeter with hay bales or plastic sheeting—low temperatures and winds howling through a narrow crawl space create a venturi effect that can dramatically reduce temperatures and freeze pipes.

• Vacation or second homes are often small, so add height. Timberframe plans make this easy to do, and the extra volume creates spacious, useful living areas and lofts with relatively little increase in cost. Also, in a vacation house, large closets are less important—save that space for living areas that you and your guests can enjoy.

• A keyless entry system that can be easily recoded eliminates the problem of extra keys for any guests, relatives, and friends who use the house in your absence, or for maintenance people or the police or fire department—who may need instant access in an

If you plan to situate your home well off the main road, consider the high cost of running utilities over long distances. Heavier and more expensive wiring has to carry adequate voltage, and per-foot costs can quickly add up. Your alternatives include installing utility poles on your property or running underground wires. Although all of this privacy can be expensive, especially if your budget is already stretched, the additional cost will be amortized over the number of years you live in your home and will pay "satisfaction" dividends for a long time to come (you can put off some expenses until later on, but you can't move your house farther back from the road once it's built). Be aware, too, that some municipalities offer subsidies to homeowners who bury utility wires on their property, which can significantly reduce expenses.

Integrating a new home within its surrounding environment also means taking into account the existing trees, rock outcroppings, wetlands, and general topography. Today, environmental laws and local review boards frequently determine what landscape features may or may not be altered when you prepare your building site.

One way to avoid problems and gain maximum value from your existing site is to hire a landscape architect to develop a comprehensive site plan. Typically, the goal is to preserve the most attractive (or legally unalterable) natural features while locating development where it creates the least environmental impact. Although this adds yet another expense, it can actually save money in the long run. A well thought-out site plan can reduce the work required by an excavating contractor, and it can prevent environmental missteps that require costly mitigation. It can also save novice homebuilders from an all-too-common mistake—moving out to the country only to re-create their former suburban or city landscape simply because they were unaware of the many options available to them.

emergency. If you plan to share your vacation home with other families or rent it part-time, build in a lockable owner's closet to store personal items, tools, and other valuables.

• As any vacation-home owner will tell you—people at their leisure take more, and longer, showers. Don't scrimp when you buy your water heater. Install the next-biggest size tank that you think you'll need. Also install an outdoor shower to keep beach sand and mud outside where it belongs.

• Appliances in seasonal homes are used infrequently—only one-sixth of the time used in a year-round home. Unless you have money to burn or expensive tastes, consider buying ordinary American-made appliances—they'll cost less, cook just as well in most cases, and last indefinitely.

• Because you won't be around your vacation home much of the time, employ building strategies that keep maintenance to a minimum. Enclose or box open eaves and cornices to discourage foraging raccoons and eliminate nesting spots for bees, bats, and birds. Also landscape around your home with zero-maintenance techniques so you won't have to work—whenever you find time for a visit.

it's different in here

**DAVIS FRAME COMPANY/
RICH FRUTCHEY**

*Energy-efficient glazing is available
in half-round and quarter-round
lights that can be assembled in
various combinations to create
custom window walls.*

**PREVIOUS PAGE LEFT:
LINDAL CEDAR HOMES, INC.**

**PREVIOUS PAGE RIGHT:
HABITAT POST & BEAM, INC.**

your home, your way

Although timberframe construction is essentially a structural system,
a building method renowned for its strength and its architectural
history, timberframe homes are really all about interior design.

That's because Timberframes, by their nature, create such interesting
and visually complex interiors. Whether the first step on the path
to the home of your dreams is a simple sketch drawn from your
imagination or you fall in love with a model home or a timberframe
showcase that you find in some magazine, the odds are high that it
is the interior of the home that captures your attention.

No matter how people and timber framing come together, there is
usually something about the interior space that strikes a resonant
chord in their hearts. To help owners and would-be owners make
the most of these wonderful and unique interiors, it's worth exam-
ining some of the basic elements common to timberframe design.

When most people think of Timberframes, the images that
spring to mind are soaring spaces supported by natural wood
beams, cathedral-like ceilings, expansive walls of glass, and great
rooms that integrate kitchens with wide-open, hearth-warmed
living areas. This is no flight of fancy—Timberframes actually do
look like this and typically feature this kind of wish-list décor.
Of course, there are no rules that dictate how each interior
should look, and manufacturers are quick to remind buyers
that Timberframes are adaptable to virtually any architectural
style. They offer enormous freedom of choice in both interior and
exterior design.

**DAVIS FRAME COMPANY/
RICH FRUTCHEY**

By leaving the risers open, this winding staircase defines separate areas of a multilevel home without making the space feel closed in or confining.

**DAVIS FRAME COMPANY/
RICH FRUTCHEY**

Interior French doors create a separate formal dining room.

The expanded view illustrates the many options Timberframe owners have available to divide space by function; among them are changing levels for the dining room, a loft library, furniture placement for the main living area, and interior French doors.

How individuals relate to living space is highly subjective. Some find large-volume interiors impersonal and overwhelming, while others approach the same expansive, wide-open areas with a sense of freedom and inspiration. On the practical side, there are people who object to the way sound travels in large rooms, or who consider large-volume interiors to be wasted space that could be better used for extra guest rooms. Others see these spaces as opportunities for furniture and furnishings that transform grand expanses into cozy, intimate, even dramatic settings—and wouldn't trade their prized vaulted ceilings for just another bedroom.

If you fit the former profile rather than the latter, you can always design your timberframe home with flat ceilings and put that extra bedroom up above. Many Timberframes, in fact, faithfully replicate authentic early American homes and inns, most of which had low-beamed ceilings, smallish rooms, and relatively few windows. As in everything, beauty is in the eye of the beholder. Your home is your choice. The trick is to make the design work for you.

Trends also change over time. Interior spaces that combine a great room or "hearth room" (formerly known as the den) with an open, eat-in kitchen is a currently popular concept. But it wasn't always. Many homebuyers, in fact, prefer a different arrangement for their homes, with smaller, more manageable living rooms, separate kitchens and dining areas, and semi-private areas such as media rooms and playrooms for family activities.

And it pays to be realistic, too. People who design that cavernous great room with a massive fireplace costing thousands of dollars, and picture themselves gathering 'round the fire with the whole family singing songs and spinning yarns night after night, or spending

DAVIS FRAME COMPANY

When the vaulted ceiling is more than 24 feet high at the ridge, hefty upholstered furniture is needed to moor the voluminous living space and anchor the décor.

romantic evenings with "just the two of us" may be fooling themselves more than a little bit. Nights like that simply don't come along as often as *Frasier* and *Monday Night Football*, not to mention back-to-school, ballet practice, and all those times when the babysitter doesn't show up.

Most timberframe manufacturers don't want homebuyers to blow their budgets on extravagant, over-the-top features they'll never use. There are enough real options—including some very expensive but more worthwhile ones—available to them if they have the money to spend. Besides, these companies know it's simply bad for business when their customers end up unhappy, broke, or both. One manufacturer urges customers to lavish a portion of their budgets on one thing that will make them truly happy in their new home. It could be a custom window wall that captures a stunning view or a world-class kitchen for people who like to cook while they entertain or maybe something as simple as a private deck with romantic French doors off the master bedroom.

The point is, Timberframes are idiosyncratic structures, and they allow for everything in home design and décor that you've already imagined, and even some things that you haven't. Knowing the difference between them and other, conventional homes is important, but once you understand where those differences lie you can use it to create a unique home that reflects your individual personality and lifestyle.

HABITAT POST & BEAM, INC.

The linear structure of this home is reinforced by the minimalist place-ment of furniture and rugs. Touches such as a landscape painting, the rich wood ceiling, and the period secretary warm the hard edges of the post-and-beam interior.

visual weight

We often think of homes in terms of square footage—how much floor space does it have? But what creates the sense of roominess may be more a function of volume. When it comes to timber framing, of course we think of open architecture and vaulted ceilings that "open up the space." One can create greater volume by adding eight, ten, or even twelve inches to the overall height of a room. This does two things: In areas where there is a vaulted ceiling, the space becomes even more voluminous and soaring. In other areas where there is a flat ceiling, such as in the kitchen or living room, the second-story loft can have a low knee wall that makes the space more useable.

After designing the interior space to suit your needs, consider the furniture—and what designers call the *visual weight* of it. "In a home with a cathedral ceiling, delicate furniture on spindly legs will actually appear to float, while more ample pieces, deep colors, natural textures, and bold patterns will anchor the interior," advises Linda Corzine of Decorating Den in Louisville, Kentucky, who has won national honors for her work on timberframe homes and is co-author of *Timberframe Interiors.*

This is just the opposite of designing furniture for those intimate nooks and crannies and loft spaces where low or pitched ceilings are the order of the day. In these tighter spots, consider furniture that is more "see-through" by using glass-top tables and upholstered pieces on legs to create an impression of open space. Ikea and Crate & Barrel, for instance, offer a wide selection of lightweight furniture.

**THISTLEWOOD TIMBER
FRAME HOMES**

*Dormers are ideal for creating
conversation areas and making
hallways, landings, and stairways
more than just pass-throughs.*

a sense of purpose

Consider the focal points of the room, too. The hearth room of a timberframe home is likely to have the most imposing handcrafted truss, whose color and mass must be taken into account when selecting furnishings. In a room with a magnificent view, large expanses of glass, specialty windows, and French doors integrate the interior with the exterior.

If you need a place to start, follow Mom's advice: Begin with the sofa and build around it, considering the activity or purpose of the space. A deep sofa that you can wallow in is great for sitting by the fire or watching football, but doesn't work well if the space is to be used for formal entertaining or community group meetings.

Above all, integrate the furnishings and decor into the architectural and structural elements, drawing cues from the wood trusses, beams, and posts to designate special-purpose areas. In many cases, structure can be used in the public areas to form two to three intimate seating areas. Add cues that define the purpose of each area. For example, position two large overstuffed comfy chairs in a corner with a small table between them, add a few books or a pair of antique glasses, and the space will declare itself a reading nook. A small desk set between posts signifies a bill-paying or writing center, and a backgammon board displayed on a table with a carafe of brandy says this area is for two people to enjoy some informal game playing.

Except in very small homes, the opportunity exists to create at least two conversation areas and perhaps a cozy reading corner that can, with the addition of a side chair, become a third conversation area. Creating these individual spaces, each with a different level of formality and scale, adds interest to the overall room, increases versatility, and makes social occasions more manageable.

Most importantly, do not overcrowd your interior—in fact, build it up over time, adding furniture and décor gradually to avoid having a large open space end up cluttered with so much stuff that no one sees the beauty of the timberframe structure.

color and texture

In addition to misconceptions about how we treat space in a timberframe home, we often fall into the path of least resistance when it comes to color and texture. All-white walls with natural timbers are *not* the only way to go. Timberframe interiors may be monochromatic with all the surfaces painted white with the play of light and shade defining the structural elements. Or natural timbers may be contrasted with highly textured painted faux finishes on the walls, ranging from pastels to deep maroons and blues. The ceiling surface, while often done in V-matched pine boards, can be done in plaster to provide a contrast of stark white against rough-hewn beams as we might see in a colonial-era home.

Consider "pickling" as a treatment for wood elements to give your home a fuller palette and greater textural variety. There are also new high-gloss finishes that work well on beams and posts in the kitchen, bath, and game room. With new interior finishes available in low VOC (volatile organic compounds), interior painting and staining is a lot quicker and easier for the do-it-yourselfer.

Other products are also available to create textured surfaces, including wallpaper that mimics "mopped" painting and other faux finishes. One advantage of wallpaper for the do-it-yourselfer is that it gives you absolute control over the outcome. It can also be completed working an hour or so in the evenings without a lot of prep or cleanup. Just size the walls and hang a few strips of prepasted paper at a time.

**BORDER OAK DESIGN &
CONSTRUCTION LIMITED**

*The yellow ochre tone of the walls
and the natural stone beautifully
highlight the rustic beams of this
home.*

Our timberframe decorating expert, Linda Corzine, offers this color
selection tip: Wood is a color, too, and is more yellow than most of
us perceive. Bring home color swatches in the yellow and ochre
ranges from the paint store and find one that matches your beams.
Later, when you select other colors for paint, wallpaper, and window
treatments, you can match them to the wood without being
distracted by its grain and texture.

THE ART IN A TIMBERFRAME HOME

One big advantage of a timberframe home is having all that open
space to display paintings, photography, and sculpture. What's
more, the vertical posts and horizontal beams can be used to define
and frame separate mini-galleries within the overall space—one
area can be dedicated to black-and-white photography, while
another can be used to display large contemporary paintings—all
within the larger open public space of your home.

In addition to painting, photography, drawings, and sculpture,
consider decorative objects as art for your home—antique quilts,
tapestries, old woodworking tools, a collection of boxes, Depression
glass, or any other object you love. These informal *objets d'art* add
texture and interest to your timberframe home and reinforce the
love of craft that brought you to timber framing in the first place.

For guidance on how to locate, buy, and display art, we went to
Renée Fotouhi. When Renée was a kid, she learned to navigate New
York galleries with her father, an inveterate collector. After earning a
degree in art history, she went to work for both Sotheby's and
Christie's, and became an international expert on Cézanne drawings.

LINDAL CEDAR HOMES, INC.

Break the rules—chandeliers are not just for dining rooms and grand foyers. A magnificent chandelier is a wonderful focal point in the main living area or great room, especially one with a vaulted ceiling.

DEFINING SPACE

For guidance on defining space in the expansive open areas of a post-and-beam home, we asked Boston interior designer Jeff Ornstein, whose experience ranges from post-and-beam and log homes to some of the world's signature hotels and resorts. He sees a connection. "A well-designed great room in a post-and-beam home is a lot like a well-designed hotel lobby in which separate groups of people can engage in intimate conversation and feel perfectly private and comfortable."

He offers this advice:

- In a Timberframe lacking a foyer, one can be defined with visual cues such as a small table for keys and mail, an umbrella stand, and a dramatic area rug.

- Don't fight the structural elements. Use the posts, beams, and trusses as natural "dividers" of individual spaces within the great room and other large areas. These spaces can then be further defined—sconces in the dining room or a ceiling fan over a conversation area are obvious examples.

From her New York gallery, Renée Fotouhi Fine Art, Ltd., she gave us this advice:

• **DEVELOP *YOUR* THEME**

I like Cézanne, but everyone doesn't have to. So don't get talked into buying a painting because someone *else* thinks you should. Art defines your home and your personality—so indulge yourself—display pieces that reflect *your* aesthetic values and taste.

• **CONSIDER THE ACTIVITY**

An energetic abstract painting may be terrific in an office tower lobby—where you want to feel the commercial energy. But that same style may overpower your home library or den. In this environment, consider something more geometric—a pair of Josef Albers paintings or prints, for example, will visually reinforce the thoughtful mood you wish to create for this space.

• **THREE DIMENSIONS**

Luckily, Timberframes have sufficient space to do justice to large sculpture. Sculpture must be installed where you and guests can comfortably view the piece from all angles. Looking at sculpture from only the front is like leaving the opera after the second act—you know how it is going to end, but the dramatic tension is never quite resolved.

• **A SENSE OF PLACE**

If you are on a budget, be resourceful. Instead of buying generic poster prints from the local frame shop, consider more creative alternatives. Display aerial views or historic maps and postcards of your new town. Purchase work by local artists that you like. Consider photos of your Timberframe being cut and assembled. Display these photos together with antique pictures of a sawmill or barn raising.

• Give every space a focal point using accents such as tapestries, distinctive wall coverings, plants, paintings, or sculpture. If the view is of a mountain range, reinforce the motif with an Ansel Adams book of photos and antique cameras on the coffee table.

• Beware of strong overhead lighting. It makes people look haggard and causes the trusses, beams, and posts to cast ominous shadows. To achieve a more pleasing effect and soften the edges of the structural elements, light fixtures must be at various heights. Use table lamps, floor lamps, sconces, pendants, and torchieres that cast upward beams. Having multiple and variable light sources enables you to create a whole range of personalities for your home—making it comfortable for every activity from children playing games on the floor to adults sipping wine by the fire.

• BEWARE OF ART EXPERTS OUTSIDE THEIR FIELD

Art dealers and gallery owners are not experts in every school of art. Beware of those who make broad, overly emphatic statements outside professional limits. Curators devote their lives to studying extremely narrow areas of art. A reputable consultant will locate the appropriate and qualified experts to evaluate and authenticate a costly work of art before you buy it.

• WHAT'S IN A NAME?

Even great artists had to pay bills. Picasso scribbled hundreds of drawings on napkins—many say as payment for restaurant tabs. That doesn't make them priceless.

• APPRECIATING ART

Stories of flea market finds turning out to be masterpieces abound. Don't count on discovering that the underpainting of a mediocre old master painting is a Raphael. Such occurrences are rare. Most important, do not buy art for appreciation; buy art you appreciate.

• DISPLAY

No matter what your tastes, a qualified art consultant can help you install and display works of art to their best advantage. The frames and matting must, of course, work individually with each piece and together with the collection. Height plays a role, too. Hang art lower in the dining room and bedroom. Consider aesthetic distances—don't put a massive piece of furniture in front of delicate drawings or etchings that will keep the viewer too far away to appreciate their mastery.

• LIGHTING YOUR COLLECTION

Even the best lighting, aimed badly, will make your collection go flat. After working on hundreds of gallery openings, it still excites me to set the lights and watch the show come alive. Ask your art dealer for help. Tip: If you have a meaningful or important painting and know where it is going to be displayed long-term, do not install track lights. Have the electrician work with your art consultant to install the proper fixtures with adjustable heads at fixed distances to eliminate the track for a cleaner, more sophisticated look.

**THISTLEWOOD TIMBER
FRAME HOMES**

*Stone columns buttress the
beefy hand-hewn wood trusses
and rafters.*

room by room:
timberframe spaces

GETTING STARTED ON THE MAIN EVENT

The living room, great room, or lately, the *hearth room* is the most
public space and most-often-seen room in the house. While your
new home may have a special-function den or entertainment room,
a smaller timberframe home may lack a "casual room." In this case,
the hearth room must have multiple personalities, functioning for
every event from quiet evenings at home to gathering around the
television for Superbowl to holidays with the family. It's the room
where memories are made.

You may want to develop an interior with multiple functions that
takes full advantage of the volume of your Timberframe, but where
do your begin? We posed the question to our expert Linda Corzine,
who told us emphatically, "Start with the sofa—this is old but sage
advice as valid now as it was when Grandma did her parlor."

While interior designers are trained to take in the big picture, others of us have to go step-by-step. To choose the right sofa for your
needs, ask yourself a few key questions. Will the living room be used
on a regular basis? Will it also be used as an entertainment area? For
family? For business socializing? "For instance, you may not care
about pleasing guests, and just want a place to wallow with your
spouse. If so, go out and buy the comfy sofa that works for you in
the color and texture that feels right," says Corzine.

THISTLEWOOD TIMBER FRAME HOMES

A truss arch frames the setting in this great room, while the tapered fireplace surround leads the eye skyward.

"Starting with the sofa," is great advice, but doing so is often daunting. Perhaps you've checked out every furniture dealer within 500 miles and can't find anything that really moves you. In that case, focus on one small key element that you know will go in the hearth room: a chair, a fabric, a color, a trim piece, a painting, or an architectural design. Having at least one item selected that *must* go into the room will give you a starting point. For example, one couple owned an antique family photograph album that they wanted to have on display. This got them thinking about matching the sofa to the colors of the book and, *voilà*, they were off and running.

The color for the walls can also determine the look of your room. For example, if you have your mind set on red walls, pick a shade and match the sofa to it. You can then again focus on the sofa as your room begins to take shape. If you choose a plain sofa, pick up just one element, such as its color or texture and use it in a patterned fabric on upholstered side chairs. Once you have laid this essential foundation, just keep building.

WOOD SIDE CHAIRS

For additional seating, consider wood. The Shakers, whose houses had a simplicity much like today's post-and-beam homes, had the right idea. They designed wood side chairs with higher backs, and upholstery on only the seat and arms. Extra chairs, like works of art, were hung on the walls—bas-relief sculpture with a purpose for the practical-minded Shakers. They could be taken down when guests arrived, yet still be out of the way the rest of the time.

Shaker chairs are available in a variety of styles and, with the addition of a toss pillow, can be made to coordinate with the upholstered pieces and overall theme of any interior.

- Dark-colored tablecloths contrast well with wood ceilings and beams and create a dramatic backdrop for bright, decorative items such as crystal and china (for the same reason jewelry and gems are often displayed against black or deep blue velvet).

- Even though the rest of your timberframe home is informal, it's really okay to have a fancy or formal mahogany table. Just buy a custom table pad to protect the surface against everyday perils.

- A round dining table feels more comfortable in a square or nearly square room. Consider benches for an informal look.

- Instead of a buffet, consider an interesting shelf of granite or old wood mounted between the posts and integrated with the frame. Such shelves are excellent for not only food service but also for lamps that complement the ceiling fixture and balance light throughout the room.

- Sconces add ambiance, but make sure bare bulbs are not visible from an upstairs loft. Covered or enclosed sconces are available for such situations.

DINING ROOMS

The pendulum of informality in timberframe interiors seems to be swinging back in time with the return of the formal dining room. Perhaps this is a reaction to 9/11, when homeowners wanted to recapture a sense of family and give Sunday dinner the special event stature it once held—and rightly deserves.

The farmer or eat-in kitchen emerged in timberframe homes in the 1970s and was perhaps part of the self-sufficiency movement—after all, the Waltons gathered in the kitchen. Timberframe homeowners today find comfort in treating the dining room décor differently, making it more formal than the rest of the house.

The dining room should evoke a feeling of family more than any other room in the house, and since form follows function, be sure to consider the food as an actual design element. This includes not only the actual food itself, but also the china, crystal, silverware, and table linens, which all play roles in setting the tone of the room. The dining room should engage all the senses, especially that of smell—a powerful sense for most of us. For instance, the aroma of cinnamon can take you instantly back to the warmth of your mother's kitchen when you watched her bake apple pies on cool fall days.

Thus, when selecting a table for the dining room consider how it will look with a range of china, from your everyday dishes to the fine bone china inherited from your grandmother. In many cases, timberframe homeowners are leaving the dining room table set all the time with large "charger plates." It looks great and also keeps family members from using the dining room table as a catchall for mail, car keys, and schoolbooks.

LINDAL CEDAR HOMES, INC.

A planked ceiling makes this formal dining room appear warmer and more intimate.

- The dining room rug is key to comfort and formality, but its pattern should not detract from the main event, which is the dining room tabletop décor. All four legs of the chair should be on the rug when people are seated at the table.

- If your dining room has a cathedral wood ceiling, with a beam running around the top of the wall, ask the carpenter to build in a strip of wood for cove lighting. It will cast indirect lighting on anyone seated at the table and make the ceiling warmer.

- Set designers know that if you want to tell the audience something important is going to happen in a room, all you have to do is hang a chandelier onstage. Likewise, if you want to make your dining room special, forget low-voltage pendants and other modern alternatives. Go with the crystal chandelier. Crystal, like diamonds, is *never* out of fashion.

Timberframe tip: The wood posts are ideal for sconces, but they must be wired ahead of time. Likewise, if at all possible, plan the dining room so that a beam cuts through the middle. A chandelier can be mounted from it without installing an extra cross member.

BELOW:
**CARGILL/BLAKE ARCHITECTS &
BUILDERS /SANDY AGRAFIOTIS**

*Natural beams coordinate with
rustic furniture while forest scenes
incorporated into the chandelier
and chair backs give the home its
sense of place.*

NEXT PAGE:
TIMBERPEG/ROGER WADE

*A wrought-iron chandelier is a
countrified counterpoint to the
formal table setting, oriental
rugs, and art.*

**THISTLEWOOD TIMBER
FRAME HOMES**

*If it is to function well, a kitchen
island needs ample room around
it—42 inches minimum, say experts.*

HEARTHSTONE, INC.

*Openness between the kitchen and
dining room can make cooking for a
dinner party a participatory event
and a highpoint of the evening's
entertainment.*

KITCHENS

A truism among space-planners and interior designers is that the less time people have, the more area they allot for kitchens. Perhaps this is because, in our frenetic, time-frazzled lives we still want to enjoy traditional pleasures without a lot of wasted time and energy. Larger kitchens with space-efficient cabinetry or pantries reduce the number of trips to the supermarket. And larger kitchens allow ample space to prepare a meal for a crowd—perhaps with the guests participating. When it comes to an intimate gourmet meal for two, the larger kitchens of today offer several food preparation areas that make it easy for couples to cook together. In many cases, very flexible, dramatic lighting is used to establish a romantic eat-in space in the timberframe kitchen that is often open to the main public space.

THISTLEWOOD TIMBER FRAME HOMES

A two-level countertop offers space for casual dining while providing a backsplash to accommodate electric outlets and task lighting.

In addition to greater space, homeowners are demanding more luxury and electronically controlled appliances that take a lot of the guesswork out of meal preparation and make cleanup a breeze. Dual dishwashers, for example, eliminate the need to stack dishes and silverware that are used daily and are boons when it comes to keeping up with glassware during a party. Professional cooktops with up to six burners and an easy-clean grill, as well as dual ovens, make it simple to prepare meals in advance. Larger refrigerators and freezers are becoming the norm in many homes today.

CABINETRY

Just because we love timber framing doesn't mean the cabinetry has to be all natural wood. In fact, contrast is the key to making the structure stand out. A popular approach—especially in a timber-frame home that's reminiscent of the nineteenth century—is to create a kitchen that appears to have evolved over the years. Your cabinets don't have to be lined up like soldiers of storage.

Consider freestanding elements such as china closets, Hoosiers (either reproduction or new), grain bins that double as recycling units or kitchen islands, and linen wardrobes. Old law office bookshelves, pie safes, or glass-front mercantile display drawers from turn-of-the-century dry goods stores are perfect storage spots for tablecloths and napkins. Incorporating these bits of the past results in a kitchen that acts contemporary yet looks like it has always been there, with pieces added when the need arose.

Searching through antique shows for individual pieces to create an eclectic look can be too daunting; fortunately, a number of kitchen cabinet companies are responding to this trend with cabinets that appear to have been pulled right out of the past. Even though they

BORDER OAK DESIGN & CONSTRUCTION LIMITED

Massive summer beams carry the floor joists in this authentic Tudor-style kitchen in which the well worn butcher block island adds loads of character.

appear to be antiques, they have all the practicality one needs, including such modern amenities as lazy Susans, wastebasket caddies, and drop-down storage in front of the kitchen sink to keep scouring pads out of sight.

Kitchen design and cabinet expert David Leonard, of the Kennebec Company in Bath, Maine, designed and built an assemblage of cabinets that re-created a handed-down, generations-old look suited for the post-and-beam Weatherbee Farm featured on the PBS series *This Old House*. For the timberframe homeowner or builder, Leonard offers this advice: Never accept useless filler spaces in a well-designed kitchen. And, along with the trend toward traditional kitchens, he notes, "Pantries are great for storing bulky items so that the homeowner can buy food in larger, more economical quantities—thus reducing shopping time from a hectic schedule. Pantry shelves are also less costly than cabinets on a per-foot-of-storage space basis."

Adds Leonard, "Remember, too, that it's not the amount of storage space but the efficiency of the kitchen that matters. Do not pay for expensive cabinets to store a chafing dish used only at Christmas, or for thirty extra pots and pans. Throw away the stuff that you won't ever use—before you start to design the cabinets."

For bulky pots and pans, large drawers are ideal for keeping cookware within reach. Leonard also points out that there are so many racks, lazy Susans, sliders, and other organizers available that it is easy to get caught up in the momentum. But to stay within budget, do not over-accessorize.

TIMBERFRAME KITCHEN TIPS

- To save time and materials, some builders install the countertop's vertical backsplash flush with the post faces, then fill in the space between counter and wall with a small shelf. Others prefer to fit the countertop material around each post for a more built-in look.

- If your Timberframe has an open plan and the cabinet tops can be seen from a loft or stairs, they should be finished with the same quality wood and color as the cabinet faces. Also consider how light sconces will look from above. Closed-top sconces are available to keep bare bulbs (and dead moths) out of view.

- Don't forget to include artwork in your kitchen. Photographs, paintings, or your favorite *objet d'art* can add an important measure of personality— just be sure they are framed to withstand the rigors of kitchen life.

TIMBERPEG/RICH FRUTCHEY

An inviting eat-in kitchen is a natural addition to a timberframe home.

- Instead of an ordinary, workmanlike exhaust hood over the range, choose one of the many design styles available. They range from sleek, contemporary stainless-steel models to more traditional ceramic-tiled hoods.

- Islands are wonderful for cooking and casual dining. The minimum space required for good traffic flow is forty-two inches all around it. But everyone has a different sense for the space that they would like. Before permanently mounting an island to the kitchen floor, live with it for a while to get a feeling of how the space will work for you.

Consider the failings of your existing kitchen and how you like to cook. If you enjoy preparing meals with your spouse or guests, consider an island or peninsula that can be reached from both sides, thus doubling the useful counter space and making it convenient for both husband and wife to work at the same time. An island with its own cooktop and countertop electric outlets for food processors, mixers, and other appliances are other conveniences that can stretch the usefulness of available work surfaces. If you do a lot of cooking and baking, think about a table in the center that allows you to work while seated. It also makes dining in the kitchen a bit more elegant than perching on stools at a raised counter. For an added touch of class, add a chandelier in a metal finish that compliments the range hood.

Task lighting may be built in under the cabinets, but add small attractive lamps that can be plugged into the counter for a homier, more decorative look.

MAKE IT FUN!
Chances are you built a Timberframe because home activities—cooking, gardening, sewing, crafts—are what make your life interesting and fun. Take a tip from professional chefs: Every chef learns that the key to preparing a successful meal quickly is *mise en place*—which means setting out items to be used in food preparation neatly and within easy reach. In professional kitchen parlance, a successful chef will be said to have "good *mise*." Even though home kitchens have hiding places for everything, do not be afraid to keep spices, non-perishable ingredients, and a selection of cooking oils with pour spouts readily available on the countertop. The look will be warmer, friendlier—and your home will have "good *mise*."

MIKE AND KAREN SOLVE A TIMBERFRAME KITCHEN PROBLEM

When designing a kitchen, which has to accommodate appliances, fixtures, and storage cabinets, working around the posts and beams can present unique challenges. A major supporting timber can't be moved or cut away like a two-by-four if it conflicts with an appliance fit.

In our house, the kitchen was a critical element in our overall design scheme—the heart as well as hearth of our home, a central focal point viewable from several rooms away, with an elaborate oak-beamed and cedar-planked ceiling done up in dramatic timberframe style. To make the most of it, we called in a kitchen design expert who developed a signature cabinet plan centered around a professional range with a large stainless steel vent hood. This hood required a tall, matching steel chimney that would extend to the ceiling and vent through the second-floor joists to an exterior wall.

On the plan, everything looked fine. The custom cabinets and vent hood were ordered well in advance of construction so they would arrive on time. We raised the frame and made the kitchen ready for their arrival.

Delivery was just days away when, while reviewing the cabinet plans to prepare for installation, we noticed that a major supporting beam was squarely in the way of the vent hood chimney. Apparently, this timber—a twelve-foot-long, six-by-nine-inch square, half-ton of solid oak that spans half the length of the room—never appeared on the kitchen designer's blueprint, and no one noticed its absence.

With the beam in the way, our custom-fabricated chimney would not reach to the upper floor joists, and therefore the critical vent run was blocked. Because the kitchen is a central room, there were no alternative exterior walls to vent through. The entire plan—the room layout, the

MICHAEL MORRIS

The range hood and its chimney were custom cut to integrate the chimney with the beam in the home of co-author Michael, and his wife, Karen.

ROCKPORT POST & BEAM/ BRIAN VANDENBRINK

Simplicity is the byword in this kitchen, which derives its beauty from the sleek vertical and horizontal planes of polished granite and copper surfaces.

placement of the range, the custom-fit cabinets—was in jeopardy.

Compromise seemed out of the question. Suddenly, a wholesale change in plans was necessary—and it had to be done quickly, or our entire building schedule would be thrown off. If the beam couldn't be moved or altered, we would have to work around it.

We called the vent hood manufacturer and ordered a new chimney to be custom-cut so that it appeared to slice through the heavy timber, but in fact wrapped around it. The vent, which now terminated just below the beam, was rerouted through the back wall of the kitchen, down to the subfloor, and out the nearest exterior wall. It took some doing, but we did it. Visitors to our home today can't see our mistake, and will never know how much effort was required to correct it.

In the end, we learned a valuable lesson about timberframe construction: What you design into your frame is what you get—permanently and immutable, for most practical purposes. If you need to make construction changes after the fact, and they involve frame alterations, they will be costly and difficult, if not impossible. So plan carefully, and try not to change your mind about where you want your post and beams later on.

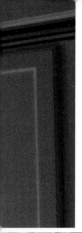

MASTER BEDROOMS

The first factor to consider must be decided during the planning stage. Do you want the bedroom to
have a voluminous vaulted ceiling or to be cozy and warm with a flat ceiling? Everyone is different, so
determine how you *feel*, even going so far as to spending a weekend in a hotel room with a cathedral
ceiling. Think it through; after all, the master bedroom is your escape from the world. Make it special for
romance, luxury, warmth, and sex appeal.

Every room needs a focal point, and the bed is the attention-getter in this room. Go lavish with bedding, especially if you want to anchor the bed in a room with a cathedral ceiling. And since most timberframe homes can accommodate an eight-foot span between posts, the headboard does not have to sit against one.

On the more luxurious end are canopy beds, which help add visual weight to the room, especially if it has a cathedral ceiling. Consider additional seating in the master bedroom, such as a chaise longue, to enhance the level of elegance and fantasy.

While rules are made to be broken, one cardinal rule of good bedroom design is to situate the headboard on the wall opposite the entryway if at all possible. That way when anyone enters, they get the full impact—the "wow factor."

Of all the rooms that resist eclecticism, the bedroom comes first. Bedroom sets are still pretty much the rule rather than the exception. While it is quicker and easier to select a set rather than individual pieces, they can limit your creativity and make a room dull. Try having some fun, instead. Change the headboard to a contrasting material such as rough-hewn boards or wrought iron to give the room an element of surprise. For an added touch, go with footboards, which are coming back into style.

Add additional interest with mirrors, and remember, night tables do not have to match. Since many timberframe homes are designed to be reminiscent of the colonial era, when closets did not yet exist, consider a huge European-style armoire that is both practical and elegant.

**THISTLEWOOD TIMBER
FRAME HOMES**

*Bunk beds are a natural addition
to a child's room in a timberframe
home.*

CHILDREN'S ROOMS

The child's room of today is the guest room or home office of tomorrow. Therefore, plan for the future when designing a child's room. Hardwire it for cable television, phones, intercoms, DSL lines, extra electrical outlets, and all the trappings of technology, while building your Timberframe. Give this room flexibility; before long the need for an intercom in the nursery will give way to the need for a computer for homework. And while it may serve as your home office or hobby room once Junior heads off to college, in later years, you may need that intercom system again— this time for your grandchildren's nursery!

BEDROOM TIPS

- A dresser is a great place for additional lighting—especially in a room with a vaulted ceiling.

- Lighting in the armoire is great for finding socks that match.

- Install dimmers on every light—even table lamps—so that you can control the mood.

- Include plants to soften corners and avoid clothing pile-ups.

- Art is hung lower and nearer to the bed.

- Nightstands are good places to display personal mementos.

- Lower the ceiling with a shelf about six feet off the floor to display rare books and memorabilia.

- High ceilings can go a shade darker than the walls.

• Window treatments need special attention.
Even if you don't *need* privacy, bedroom
windows without drapery or shutters
become cold black holes at night.

BATHROOMS

Bathrooms are more spacious and more hi-tech
than ever. Gen Xers and Baby Boomers want
bigger tubs, some with contoured headrests and
portable aroma spa therapy for luxury, relaxation,
and recreation. Dual showerheads, body sprays,
and whirlpools continue to be frequently asked-
for luxuries in bathrooms, while stereos and TVs
are also gaining in popularity, according to the
National Kitchen & Bath Association (NKBA), who
will put homeowners in touch with kitchen and
bath professionals in their area.

Universally designed (also called barrier-free or
user-friendly) bathrooms are hot, too. If you are
building with retirement in mind and want to
remain in your home as long as possible, plan for
a bath that will be safe and functional should
physical capability decline over the years. Multiple
showerheads with a range of sprays from massage to pulsating, oversized shower benches,
grab bars, extended lavatories, hand-held showers,
heated toilet seats, grip rails, open seating in
showers, and safety rails are all combinations of
safety and convenience that the NKBA suggests
homeowners consider.

Looking forward, this authority on bath and
kitchen design and technology predicts a future
when it will be common to call your home from
the car, enter a personalized code, and have your
bathtub fill to a preprogrammed temperature
and level.

Yet, while bigger, more lavish bathrooms are in,
they may be costly luxuries. Of course, some
homeowners will get a lot of use out of theirs—
statistics do say we are taking longer baths. But
for most people, that eighteen-jet massage spa,
big enough to swim laps in, will get tons of use
for about a week, and then the fantasy of luxuri-
ating will evaporate with the pressures of getting
the kids to soccer. When designing your plan,
include only the features you really need.

Whether you are going small or large, lo-tech or
hi-tech, designing a timberframe bathroom
poses particular challenges. While the wood in a
timberframe home is the interior focal point,
natural wood posts and beams may be problematic in a bathroom because of moisture
buildup. From an aesthetic point of view, the
smooth, water-resistant surfaces of a bathroom
dominate the interiorscape and present too
much of a contrast to the wood. In some cases,
stone and other natural materials that have
some texture can bridge the gap between, say, a
slick acrylic tub and the wood beams.

In other cases, one may want to eliminate any hint of the post and
beam structure in the bath by boxing in the beams to create an
environment that is strictly twenty-first century. Choose the approach
that satisfies you.

LAUNDRY ROOMS

The bedroom/bathroom area can often be a practical place to put
a laundry room, since this is where most of the dirty clothes come
from. If you consider a bathroom laundry room you will not only
require space for the washer and dryer, but also for folding, supply
storage, soaps, softeners, and possibly some ironing facility.

Several companies offer narrow stacking washers and dryers that
reduce the amount of floor space needed. Those that are twenty-
seven inches in width and depth are fairly typical, while a few
manufacturers offer stacked washers and dryers that are even narrower.
Look for a washer that has an automatic shut off in the event that a
hose springs a leak. Insist on shut-off valves on all of the supply lines
in the bathroom. Stainless steel–wrapped high-pressure hoses offer
an extra element of safety against flood damage.

Be sure to increase the capacity of the ventilation system in a bathroom that will contain a laundry. Recycling moist air into the dryer
just slows the drying process. Vent straight to the outdoors wherever
possible—the more bends in the vent, the slower the drying time.

Consider the aesthetic impact when it comes to keeping deter-
gents and other laundry supplies out of view. Bifold doors often
look flimsy. Pocket doors, however, will conceal the laundry area and
are more attractive. Install doors that match the vanity cabinets to
create a unified look.

WHAT ABOUT COST?

According to the National Kitchen
& Bath Association, most kitchen
designers charge fees in the $350
to $600 range, while others base
their fees on a percentage of the
job—usually 6 percent. Whether
your job is high-end or low-budget,
it's best to shop around to find
the right designer for your tastes.
And, since you and your kitchen
will be spending a lot of time
together, look for a cabinet
company where you feel the
right chemistry.

One of the best ways to get started
is to contact the NKBA at 800-FOR
NKBA. Ask for its helpful publica-
tion, "The Little Book of Kitchen
and Bath Wisdom," and a list of
manufacturers and retail members
of the association.

THISTLEWOOD TIMBER FRAME HOMES

Beautifully crafted built-in furniture with personal touches make a home unique, as demonstrated in this vanity.

THE NKBA'S PREDICTIONS

Here is what the NKBA sees over the horizon for today's bathrooms:

- Towel warmers to take away the chill.

- Fog-free medicine cabinet mirrors with electric defogging devices.

- Toilet technology with ventilation systems that eliminate 98 percent of toilet odor.

- Marble, granite, and artistic ceramic tile that turns the shower into a grotto.

- Faucets, the jewelry of the bath, off-set with chrome, two-toned metal finishes, fourteen-karat gold, or stainless steel, with the option of retractable or electronic hands-free faucets.

- Shower doors define one of the greatest changes in millennium baths. New doors are manufactured in powder-coated finishes, with rounded curves in many different colors.

**EUCLID TIMBER FRAMES/
JOHN PHIPPEN**

*Any decorating style can work in
a timberframe home. In this bath,
dark carved wood creates a
Victorian formality.*

- Glass everywhere: glass sinks,
  glass countertops, and glass
  block in the basic green, plus
  new shades, such as cobalt
  blue and yellow.

- Multicolor bathroom décor
  in earth tones that are
  punctuated with brilliant
  contrasting primary colors.

- Adjustable lighting to fit any
  occasion, from getting ready
  for a power breakfast to a little
  night romance.

**EUCLID TIMBER FRAMES/
JOHN PHIPPEN**

*The intricately carved fireplace
surround adds dimensionality and
drama to create a focal point for
this room.*

we gather together

FIREPLACES AND MANTELS

You have a great hammerbeam truss that cost thousands of dollars and is the focal point of your main living area. Can the fireplace be just an afterthought? We think not—especially when one considers that, after man discovered fire was a better way to keep warm and make food tasty, he also discovered a way to bring his fire indoors. His indoor hearth was nothing more than a hole in the roof—but it was the first conversation pit long before it became an interior decorators' term of the 1960s.

The benefits of central heating aside, no one has ever come up with a more inviting place to gather together than by an open fire. Perhaps this is the essence of our love affair with the fireplace and our desire to live in a timberframe environment that uses construction techniques that are more than 400 years old.

Like everything else, technology has made fireplaces more efficient, more varied, more cost-effective, and more convenient. While there is a certain joy to building one's fireplace from locally gathered river rocks and finding that special mason who can squeeze you in, it is not always possible. Today's fireplaces can be built of stone and brick imported from anywhere in the world. There are even lightweight man-made stone facings in just about every color and hue that defy detection as fakes. For traditionalists, the stately Rumford fireplace of our colonial era is coming back and has an individuality that is hard to beat.

While the old adage, "The man who cuts his own wood warms himself twice," still rings true today, a busy schedule may put a crimp on your escapades as a weekend Paul Bunyon. For those who want the ambiance of an open fire at the turn of a switch, consider a gas unit. Those that are billed as direct-vent units eliminate the cost of a flue and chimney, cutting the cost to the bare bones. Ventless gas fireplaces are also available and can be located on an inside bedroom wall so that you can wake up to instant ambiance.

**CLASSIC POST & BEAM/
SANDY AGRAFIOTIS**

*A central fireplace can be used to
divide one large area into two or
more distinct spaces, as in this
kitchen and dining room with separate sitting nook off to one side.*

fireplace options:
build it or install it

TALL, SHALLOW RUMFORDS

The Rumford made quite an impact during the colonial era, when
Benjamin Thompson, dubbed the Count Rumford by the Bavarian
Government, invented the efficient fireplace that bears his name.

Rumford fireplaces were common from 1796 until about 1850. Thomas
Jefferson had them built at Monticello, and Henry David Thoreau listed
them among the modern conveniences that everyone took for
granted. There are still many original Rumford fireplaces throughout
the country, often buried behind newer fireplace renovations.

Today, Rumfords are enjoying a renaissance, thanks to enthusiastic
fireplace designers and builders such as Jim Buckley of Buckley
Rumford Company in Port Townsend, Washington, who has thor-
oughly researched their history. Component manufacturers and
architects are also fueling interest in the Rumford.

Explains Buckley, "Rumford technology relies on a taller, shallower
firebox that radiates more heat into the room and a curved masonry
venturi installed above the fireplace opening in the throat." For
those who remember seventh-grade science, a venturi forces
airstreams—or smoke—to move faster, which in turn lowers pressure,
to make the air lighter and allows it to exit more efficiently. The
taller firebox allows one to build a teepee-shaped fire, which is
both efficient and reminiscent of the open campfire, adding to the
desirability of the fireplace as a gathering spot.

DECK HOUSE, INC.

Simplicity of form and minimal use of wood emphasize this room's dramatic focal point—a tall, multi-color brick hearth.

"One of the things I enjoy about the Rumford's taller firebox is that you not only get to see the full flame but also the smoke as it trails upward, creating a campfire effect," says Rex Hohlbein, AIA, who has designed numerous homes in the Pacific Northwest.

Given adequate ceiling height, a conventional fireplace can be converted to a Rumford or, as enthusiasts say, "Rumfordized." Prices for the design, materials, and construction of a Rumford can range from a low of $3,000 to as much as $10,000 for one that is large with elaborate stone and masonry work. Like other fireplaces, Rumfords can use lightweight stone facing to reduce construction costs.

Size-for-size, Rumfords cost about 10 to 20 percent more than other site-built fireplaces including those using a factory-engineered and manufactured firebox, as well as direct vent and vent-free gas-log fireplace units.

"Whether you turn to an architect or mason to design and build your fireplace, make sure they have fireplace experience and get references of people for whom they have designed smoke-free functioning fireplaces," advises Hohlbein. "Remember, too, that no matter how experienced these individuals may be, a fireplace may not function properly for reasons that have nothing to do with design and construction, such as prevailing winds, air pressure inside the house, and other factors."

FACTORY-BUILT CONVENIENCE

Dozens of companies around the United States manufacture cast metal fireplace units and flues that can burn gas, pellets, and oil as well as wood. Among the more commonly known names is Heatilator, a division of HON industries.

According to the Hearth, Patio & Barbecue Association, (HPBA), a factory-built unit is actually a firebox enclosed within a steel cabinet that is designed to force a blanket of insulating air that keeps the outer wall cool. Meanwhile, spacers surrounding the cabinet allow it to be installed in close proximity to the wood framing (hence the name "zero-clearance fireplace"). These units pass rigorous testing standards established by the Underwriters Laboratories and the American Gas Association and have an excellent safety record.

If you have an existing fireplace that has deteriorated over the years, the HPBA recommends metal fireplace inserts. These are heating units designed to retrofit into an existing masonry or factory-built fireplace and will burn wood, gas, or wood pellets and offer superior efficiency. Inserts can be hooked up into the existing

THISTLEWOOD TIMBER FRAME HOMES

A massive hearth, fashioned from lightweight manufactured stone with a factory-made fireplace, was easier and less costly to build than it appears.

chimney, "though a flue liner or other modification may be necessary," says the HPBA. It also notes that "vent-free inserts require no chimney or flue modification. Most have blowers to circulate the heat…[and] are used to change an existing non-efficient fireplace into an efficient, heat-producing zone heater." Both inserts for retrofitting an older fireplace and new factory-built units are relatively inexpensive—between $500 and $2,000. Their light weight eliminates the need for additional footing or other structural support. Such manufactured units in no way limit choice when it comes to the fireplace surround, mantel, and facing.

Also available are kits to build what we call a "traditional" fireplace and what the HPBA terms a "masonry heating unit." The advantages of working with a pre-designed kit are many: saving time on site, having more variety than most local masons' imaginations and skills can provide, choosing distinctive materials, and utilizing a firebox that is engineered and tested by experts. Numerous manufacturers offer a range of styles, including importers such as, Tulikivi U.S. Inc., a Finnish manufacturer of soapstone fireplaces that installs its units through a network of authorized dealers.

According to the HPBA, because they have fewer limitations than site-built masonry fireplaces, they can be easily and safely installed in almost any room. They are available in a range of sizes and can be open on one, two, or four sides.

Yet, no matter how the fireplace gets into a room, accessories, from screens, glass doors, and tools sets, designate the hearth as a special place.

TIMBERPEG/RICH FRUTCHEY

*This site-built masonry fireplace
makes good use of manufactured
stone, available in many styles
and colors to complement any
home's decor.*

THE GAS-FIRED OPTION

While the romance of chopping wood and building a perfect fire
may be appealing, a gas fireplace is the ideal solution for those with
busy schedules. It can be turned on upon arrival and off when you
leave, without any concern that embers will start a fire.

Vent-free gas units that are based on high fuel efficiency (99.9 percent), gas log sets, or direct vent units make it possible to install a
fireplace anywhere without a flue or chimney. The only precaution
is to make sure there is adequate air for combustion. Also, vent-free
units do not work well above altitudes of 5,000 feet.

When installing a manufactured fireplace, follow the advice of the
manufacturer because many local masons are not aware of the minimal clearances needed and may tend to overbuild the fireplace,
adding weight, cost, and complexity. It really is more of a carpentry
job, with an experienced plumber or technician to install the gas
lines correctly.

HEARTH AND MANTEL OPTIONS

Among the modern materials that have simplified fireplace construction is Cultured Stone, a trademark of Owens Corning. Though
there are several competing suppliers of manufactured stone,
Owens Corning claims it is the largest in the industry. Cultured Stone
products mimic the look of natural stone or brick and can be used in
a wide range of applications, including surrounds for manufactured
fireboxes and flues and Rumfords.

Introduced in 1962, and trademarked ten years later, Cultured Stone
products are made from natural ingredients including Portland
cement, lightweight aggregates, and iron-oxide pigments to create

**CULTURED STONE®—
A DIVISION OF OWENS CORNING**

*Another example of manufactured
stone, this stunning hearth
surround perfectly mimics natural
ledgestone but at a fraction of
the cost of real cut stone.*

a wide range of stone and brick textures, colors, and shapes for fire-place surrounds, chimneys, and other applications. Products include, for example, river rocks, cobblestones, brick, pavers, and fieldstone. Manufactured capstones, sills, and hearthstones in complementary colors also are available.

Veneers are cast in molds taken from natural stones that have been selected for size, shape, and texture. Each color and texture has its own blend of ingredients, and each replicates the original with a level of detail that mimics the look and feel of natural stone. Because of their light weight, the cost of building additional foot-ings or wall ties is eliminated.

No matter what option one chooses—impressive Rumford or manu-factured unit—make your hearth uniquely yours. Leave some funds in your budget to add individual touches, such as installing a laser-cut stone with a significant date inscribed or an element that reflects the locale such as a lighthouse or pine tree. Or just have guests contribute a rock, a seashell, or interesting driftwood. When enough pieces have been collected, have your mason build them into the fireplace surround—that way you'll feel the presence of every guest who ever graced your home. Don't limit yourself to masonry. Just about anything can be embedded—even bits of crockery, a dish from Grandmother's china, or colored glass. Bronze or brass bas-relief objects that have special meaning to you can also be worked into the fireplace.

For the crowning touch, come up with a unique mantel. Be on the lookout for an old beam from that barn that is being razed, a granite step, a log that washes up on shore, or even the gunwale from an old rowboat. All such items can be used to create a mantel that reflects the love of natural materials that is the essence of the timberframe home.

Let your lighting make a statement! Note how the exterior fixtures wash the walls in light and the interior lights offer a peek at the post-and-beam interior.

interior lighting

A CRITICAL TIMBERFRAME ELEMENT

Ask a dozen people what is the most important décor element and you will get twelve different answers, from the sofa to the window treatments to one artist who said, "start with the paintings and then design around them."

In reality, lighting is the king of décor. It is the element you change daily and the element that changes the way you look and feel. Savvy restaurant owners know lighting effects mood and that your date—and hence your experience—will be more romantic under low lights shining up rather than harsh overhead spots. You can be just as savvy when lighting your home.

Be aware that the additional volume in a Timberframe presents a lighting challenge. Vertical posts in an open space must be in concert with the lighting scheme, too. There are many options available, as Jeff Ornstein, founder of J/Brice Design International, in Boston, Massachusetts, explains: "The range of today's lighting options— from low-voltage systems to pendants to recessed lighting to torchieres, sconces, and more, afford maximum flexibility and give you control over the character and tone of every room in your home. Even the bath can have lighting that is adaptable to every occasion. Set one way, the lighting energizes you while getting ready for a hectic day at the office. Set another way, the lighting melts away your cares while getting ready for an evening of romance."

For really effective lighting Ornstein offers the cardinal rule: "The more sources and direction of the light, the more you can envelop

yourself, your family, and your guests in an infinite variety of light to create any ambiance you desire."

Mary McDonagh of Chimera, a Boston lighting company, adds, "To help with the more complex lighting options of today, specialized firms are cropping up with experts on staff who will review blueprints; design total residential lighting systems; supply all the behind-the-scenes wiring, tracks, and transformers; and sell lighting fixtures that are far more distinctive than those produced for the mass market."

If working with a lighting designer is not in your budget, there are sources available to help the do-it-yourselfer. Sea Gull Lighting in Riverside, New Jersey, a major manufacturer of lighting and fixtures, publishes catalogs that contain a wealth of information on equipment and technology—visit the company's Web site at www.seagulllighting.com.

Sea Gull's catalog on its Ambiance Low Voltage Lighting system covers terminology, materials, and accessories needed for track, pendant, linear, and disc lighting, the latter of which offers all the benefits of recessed lighting but with low-profile fixtures suitable for surface mounting—ideal for log home cathedral ceilings!

Another great source on contemporary lighting is GE's Virtual House Tour. The House, built in Brecksville, Ohio, near Cleveland was featured in the Pella Windows & Doors Homearama 2000. Executed by Kathy Presciano, lighting specialist with GE's prestigious Lighting Institute at Nela Park, the virtual home is invaluable for planning a lighting scheme and learning about available options. It may be toured at www.gelighting.com/virtualhouse/home.html.

BUDGET SUFFICIENTLY

Great lighting comes at a price. When planning your timberframe home beware of the innocent looking "builder's standard" allocation to cover the cost of lighting fixtures. These budgets are typically way behind the times when one considers that a simple arrangement of three low-voltage pendants over a counter can cost $3,000 for wiring, transformer, and fixtures.

In addition to covering the cost of modern lighting, make sure there is enough in the budget for some of the handcrafted, traditional fixtures that can make your Timberframe more inviting and fun. For example, Hutton Metalcrafts in Pocono Pines, Pennsylvania, has been making brass, copper, and pewter lanterns for indoor and outdoor lighting since 1965. They know that log homeowners want beautifully handcrafted products. As Tom Hutton, president, tells it, "The

**EUCLID TIMBER FRAMES/
JOHN PHIPPEN**

*A wagon wheel chandelier is the
crowning touch to this voluminous
great room, which boasts a stone
floor, bas-relief mantel, and other
rustic details.*

company was born when my father Ray Hutton built a log cabin in Pocono Pines and could not find lighting to match the rustic quality of the cabin, so he made his own. At the time, he was a food broker in Philadelphia and eventually moved out to the country to found Hutton Metalcrafts."

For a Western look, consider rustic lodge-styled fixtures by Avalanche Ranch Light Company in Bellingham, Washington. The company makes hand-hammered fixtures whose themes feature pine trees, bears, and salmon in rust, green, and seafoam patinas. Likewise, M. Star Antler Designs in Lake Isabella, California, and Arte De Mexico of Burbank, California, offer lighting fixture lines that are in harmony with the local environment.

While it may be ideal to have every lighting fixture and lamp be a conversation piece, your budget may only allow for one or two extraordinary fixtures—perhaps one in the dining room and one for the foyer where permanent fixtures play a key role in setting the overall character of your log home. Although you may be limited in the number and quality of fixtures you can afford at move-in time, it does make sense to install the necessary wiring and outlets so that

the lighting system can be expanded over time. And while you are at it, don't forget wiring for cable TV and a high speed Internet access network for all your home computers.

Building codes in most parts of the country require wall outlets every six feet. While this may seem adequate at first, there are spots near stereos, home office computers, and entertainment centers where four plug outlets or outlet strips are a must. In areas where you even think you might place furniture, or need fixtures, have the electrician install electrical boxes and cap them for possible use later.

In the open architecture of a timberframe home's interior, furniture groupings often define living areas. Consider floor outlets at strategic locations, such as where end tables are likely to be placed. Attractive outlets are available with brass covers, so that they can be obscured and safely covered when not in use.

With an adequate budget, advance planning, and an appreciation for lighting as an integral part of your décor, you can create an environment that makes staying home the best place to be.

THREE TIMBERFRAME LIGHTING TIPS

1. Install wiring for dining room and living room sconces, even if you are not considering them when you build; you may want to add them later.

2. Install lighting midway up stairways and in the middle of long hallways to maintain a pleasant feeling and increase their usefulness as art and photo galleries. Install low-level recessed lighting along the entire stairway. It's great for getting around on early winter mornings without disturbing family slugabeds.

3. Install two widely spaced electrical outlets over the mantle so that fixtures or lamps may be situated to properly illuminate artwork. Install outlets behind bookshelves where a stereo might go to avoid running cords to the closest wall outlet.

JOSIAH R. COPPERSMITH

Sconces are ideal in a timberframe home, but make sure the view from above is attractive.

THISTLEWOOD TIMBER FRAME HOMES

Discreet overhead lamps combine with a sunny window to provide ample light for this loft office.

FLUORESCENT LIGHTING—THAT IS THE QUESTION

What about fluorescent lighting? It can save a bundle on energy costs, but many homeowners remain resistant, despite advances in the technology and fluorescents that can go into ordinary sockets on table and floor lamps and fixtures.

Specify fluorescent lighting with electronic ballast, which operates the lamp at higher frequencies while producing the same light output and consumes about 25 percent less wattage. As a bonus, electronic ballasts are lighter, produce less audible sound, and eliminate flicker.

While fluorescent lamps deliver greater light output for their wattage, incandescents produce warm rather than full-spectrum light, which tends to be more comforting and flattering to facial tones. For home offices and hobby rooms, balancing fluorescent lighting with incandescent task lamps works nicely and seems to create an energizing environment. Some people claim that warm incandescent lights in their office—whether at home or work—look very luxurious but make them want to take naps.

When arranging task lighting for hobbies and home offices, a good rule of thumb is to have about three times more light on work surfaces than the rest of the room.

There are many tricks of the trade that the timberframe homeowner can employ. Indirect lighting from beams near the ceiling level can provide the general lighting in bedrooms without the harshness of the old-fashioned central ceiling fixture that is only turned on when vacuuming or matching socks.

light ideas from one of america's most noted authorities

Kathy Presciano has more than twenty years of interior design experience and has been a lighting specialist with GE's Lighting Institute at Nela Park for more than six years. She has worked on home projects across North America, designed the National Christmas Tree in Washington, D.C., and was responsible for the lighting design of the Virtual House featured in the Pella Windows & Doors Homearama 2000.

Here are some of the room-by-room lighting techniques Kathy offers in the Virtual Home Tour.

FOYER AND STAIRS

Vertical Uplighting. Grazing the wall from the bottom up using concealed fluorescent or incandescent floodlights adds depth to the space and highlights architectural features such as the multi-level detail on the stairs.

FAMILY ROOM

Portrait Lighting. Portrait lighting creates a focal point in a room, with either soft, candlelike effects or dramatic highlights. Pre-planning is key. Lamps should be selected with an appropriate beam spread for the size of the object to be lighted. Adjustable fixtures are important for precise aiming. Artworks with larger frame widths often require the light fixture to be positioned farther away from the wall.

CLASSIC POST & BEAM

Natural lighting can make even difficult spaces more attractive. Here, skylights help transform a gable loft bedroom into an intimate, comfortable retreat.

HEARTH ROOM

Cove Lighting. Concealing a series of low wattage incandescent lamps in the "cove" ceiling recess produces a soft glow of light for a warm nighttime background and complements the light of the pendant fixture.

KITCHEN

Undercabinent Lighting. Incandescent or fluorescent light sources should be positioned on the front edge of the upper cabinet and effectively shielded from view.

Interior Cabinet Lighting. Small halogen lights inside the cabinets with glass shelves highlights decorative glassware and ceramic serving pieces. GE's Profile Pucks are ideal for this application.

DINING ROOM

Mini-spot Lighting. Grazing architectural elements, such as pillars and posts, brings out texture and form to add drama to a room. Typical light sources are MR 16s, which provide a small, intense narrow beam. Light fixtures need to be placed close enough to the architectural detail, allowing light to graze the edge of the element, and fixtures should have some flexibility for easy aiming.

HOME OFFICE

Direct/Indirect Lighting. A major requirement of today's home office is the reduction of glare and reflected light on computer screens and work surfaces. Direct/indirect lighting fixtures use the ceiling as a reflector, directing light in an upward direction and reflecting it back down to deliver soft, diffuse light, as well as delivering light in a downward, more direct, position.

MASTER BEDROOM

Reading Spotlight. Bedside table lights are the most common choice for reading but can often disturb a partner trying to sleep. Two MR 16 adjustable mini-spots can be installed with an extremely tight beam to illuminate just the reading surface on either side of the bed.

MASTER BATHROOM

Mirror Lighting. Lights positioned directly over the head will cast shadows on the face unless complimented with light from a secondary source. Wall sconces positioned on either side of the mirror will help to bathe the face in even light. Bracket lights over the mirror at a height of forty-two inches above the counter can be effective.

what are some basic strategies for lighting design?

A NEW IDEA

GE makes shatter-resistant Teflon coated incandescent bulbs for table lamps that add an extra measure of safety in children's rooms. Flame-shaped decorative bulbs will bring the romantic glow of candlelight to traditional wall sconces. For crisp white, use halogen floodlights and spotlights in ceiling fixtures and track lighting. "Also look for GE's Reveal bulbs, which filter out yellow rays that normally comes from standard incandescents. Reveal accentuates the colors and textures of your log home's wood interior to really make the architectural details pop," notes Karen Farwell, spokesperson at GE's Nela Park.

Halogen floodlights are especially effective for outdoor lighting. They lend a feeling of safety, while their crisp white light dramatically accents landscape and architectural features.

In the bathroom, color-enhanced fluorescent tubes or decorative globes around the vanity mirror will spread strong, shadowless light across the face for grooming

Light plays an essential role in our ability to comprehend the world around us. Lighting systems play critical roles in how we perceive spaces and can even influence how we act in them. Lighting affects performance, mood, morale, safety, security, and decision-making.

In addition to providing lighting that is adequate for specific tasks, is safe, and makes you feel comfortable, Sea Gull Lighting professionals suggest several ways to go beyond the basics. They offer ideas worth considering that will enhance the beauty of your timberframe interior. It starts with controlling the direction of light. To explain this approach, let us begin with one object, such as a piece of living room sculpture.

KEY LIGHT

When you shine a light on an object from a single point of light, it is called "key light." It highlights contours on the object and creates shadows; the exact effect depends on the angle of the beam of light. Most of the time, an object should be lit so that you can see its front. In this case, the light source should be placed in front of and to the side of the object at a forty-five-degree angle.

FILL LIGHT

Fill light creates drama. It can either be directional or diffused. Shine a directional light on the object from the opposite direction of the key light, softening or eliminating shadows depending on the

strength of the fill light relative to the strength of the key light. Fill light sources behind the object will light the entire room evenly.

SILHOUETTING

Suppose you want to emphasize the shape of the object. In this event, soften or even eliminate the key light and directional fill light, and instead provide only fill light, either intense or diffused, depending on the clarity of the silhouette and the drama you want to produce.

UPLIGHTING

The effect of uplighting is either very desirable or very undesirable because it is unusual. Its effects range from intimate to eerie. Landscape lighting often includes uplighting to accentuate bushes and trees.

and applying makeup. To discreetly light the entire room, try low-wattage reflector bulbs in down-light fixtures or color-enhanced fluorescents to flatter skin tones.

Our timberframe decorating expert Linda Corzine offers this advice: "The guest powder room should have a warm, flattering glow that is virtually identical to the lighting used in the living areas—that way make-up applied in the powder room will work when the guest returns to the' party. Try soft pink incandescents placed in sconces and ceiling lights controlled by dimmer switches."

TIMBERPEG/RICH FRUTCHEY

Well-placed track and task lighting can be used to isolate and optimize workspaces, display artwork, or highlight intricate Timberframe details.

SPARKLE AND GLITTER EFFECTS

To add an atmosphere of elegance, add little lighting points of interest in the form of sparkle or glitter. Sharp reflections on specular surfaces in the room, such as silverware, produces sparkle. If the light source itself is the focus, such as with a chandelier, you will create a glitter effect. Beware of glare in such cases.

GRAZING AND WASHING SURFACES

On timberframe walls or on the surfaces of objects, you can change the way light impacts them so that different effects are produced. Suppose you have a wall with a rough texture that you want to emphasize. Graze the surface with light so that it strikes the surface at a sharp angle. In this case, the light source should be mounted close to the wall. Now suppose the wall is smoother, and you want to emphasize that smoothness. Wash the surface with light so that it strikes it at a wider angle.

special places for special times

NOOKS, CRANNIES, AND QUIET CORNERS

Remember when you were growing up and finally got a room of your own? Perhaps it was when an older sibling married or went off to college. All that privacy was marvelous, but before long, you were back at the dining room table doing your homework, simply because you wanted to know if anything interesting was going on. You derived energy from the comings and goings of other family members.

*Window seats make the most of
a cozy niche, providing space to
display collections, store blankets
or toys, or just enjoy the view.*

Timberframe homes offer an ideal compromise between the public and the private. Their interiors lend themselves well to creating lots of semi-private spaces throughout the house. For example, an ample landing overlooking another part of the home can accommodate an overstuffed armchair, small table, and lamp. You can create nooks that are cozy, but not isolated.

LOFTY VIEWS

Lofts are ideal spaces for kids, who can retreat to them for privacy while still keeping an eye on what is going on in the rest of the house. Lofts are also great computer centers.

WINDOW SEATS

Because timber framing lends itself to rooflines with tall, peaked gables and dormers that are often used to bring light and ventilation to the interior attic or loft spaces they create, window seats are clever ways to fill in the niches that naturally result under the eaves. A simple bench below a window is an easy project, and one that can be done by the homeowner after the room is finished. Add a section of paneling in front and a hinge for the bench top, and you have a handy storage bin for blankets, bed linens, or toys and dolls. A cushion can be purchased or custom made to fit the bench. Before you know it, you'll have the perfect spot to spend a rainy afternoon curled up with a good book.

ENTRYWAYS, HALLWAYS, AND FOYERS

The entryway of a Timberframe is often not a distinct foyer. To define this area, lay down a magnificent carpet, add a table for dropping keys or an umbrella stand that is filled with antique canes and umbrellas and you've created a space that is both useful and artful.

When decorating the foyer think about how it will be used. Aside from greeting and saying good-bye to guests, will it be used to pile the mail, hang coats, store boots, and primp before a mirror? A front hall closet can hide all coats, mittens, hats, and boots. A bench covered with a comfortable cushion can serve as a place to read mail and remove shoes. Make the foyer a place to linger, rather than pass through, for this is where guests often get their first and last impression of your home and your family.

Designers have learned, for example, that long interior views give rooms the impression of spaciousness. This doesn't mean that every room has to be exaggeratedly long. A pass-through opening between a kitchen and dining area can accomplish the same goal: making two separate spaces seem linked, and therefore larger than the individual rooms. Sliding screens, drop-shades, or bi-fold shutters can be employed as needed, without permanently sacrificing the impression of timberframe spaciousness.

LEFT:

**LINDAL CEDAR HOMES, INC./
STARK IRELAND ARCHITECTS**

*Hand-carved and wrought-iron
balusters alternate on this elegant
staircase, which has windows
positioned high and low to add
natural daylighting.*

TIMBERCRAFT

*A well-placed window sets the
scene as it floods this out-of-the
way corner with light.*

Hallways are another way to create long views, and they have the practical advantage of linking rooms and the different areas of a home. Long, barren halls without natural light have little appeal, however, especially if they dead-end at a cul-de-sac or wall. At the very least, every hallway should have a focal point at each end—a window or a table with a vase of flowers or a mirror above it. After all, what good is a long view without something to appreciate at the end of it!

Mirrors on the walls opposite windows effectively double the natural light coming into the room and make the areas appear more spacious. If the room is roofed, skylights can be used in place of windows on the opposite wall and have the same effect. Solar "tubes" or "tunnels" with highly reflective interior walls can also be used to channel natural light into interior areas far from the exterior. In much the same way, interior wall openings and even interior windows between rooms allow daylight to spread throughout a home, or even provide a surprising glimpse of blue sky or trees in places where one might not expect it.

CHAPTER FOUR
practical matters

the view from within

WINDOWS AND DOORS

Although it is not always possible, the most successful timberframe designs hint at their post-and-beam construction before one enters the home.

The adobe homes of the Southwest, for example, carry vigas (roof support beams or purlins) through the exterior walls and, in so doing, integrate the exterior and interior aesthetics. A gable-end front entryway, supported by an exposed truss will also achieve the same effect. It gives the visitor an initial clue of the timberframe within, and that entry theme will likely be carried throughout in the structural framework visible in the interior.

In much the same way, windows perform a similar function by integrating the interior and exterior. A home with large groupings of windows and plenty of glass helps the onlooker form a mental image of the interior space. Likewise, when the overall design theme and wood species of the window frames are well coordinated with the design and species of the beams and posts, the home is architecturally unified.

The options available from nationally branded window manufacturers today are virtually unlimited. But there was a time, less than a generation ago, when the selection of styles was restricted to what could be produced—or imagined—by your nearest millwork shop. If you wanted something special, such as a half-round or round-window, you paid to have it created by a local artisan. Today, the national companies offer windows in nearly every configuration, every shape, and every architectural style, as well as custom windows with highly energy-efficient glazing.

DIVIDE AND...ARGUE

Windows, of course, regulate how much natural light actually enters a house. But they also provide an aesthetic counterpoint to otherwise blank walls. How should they be arranged? Traditionally in timberframe buildings, windows were placed near the center of a wall, an acceptably symmetrical arrangement but also necessary because the post and one or two flying braces at each end of the wall precluded most other options. Moving the windows in from the corners avoids conflicts with the braces, but it doesn't do much for interior daylighting, which is another important consideration. One large window creates a very bright area that makes other areas appear dark, while several evenly spaced, medium-sized windows provide better light throughout a room.

Muntins also affect interior light quality. Historically, a framework of interconnected wood muntins was needed because glass was difficult to make in large sheets. This arrangement allowed many small panes to be used instead. But the muntins have an added benefit—the multiple small pieces of glass reflect light in all directions and help to even out the light within a room. There are those who argue that a window with muntins provides a greater amount of daylight than an undivided single-pane window of equal size but with an effectively larger glass area. If you follow this logic, you can buy windows with muntins—which are generally priced higher—and offset your cost by ordering smaller windows. To gain the full light-reflecting advantage, however, the windows must have true muntins, not the "artificial" snap-in or surface-applied grilles. It's a theory, at any rate.

There is also a longstanding debate between those who insist on true divided lights and those who favor removable grilles. Both

**BORDER OAK DESIGN &
CONSTRUCTION LIMITED**

*Authentic divided-light windows
use lead caming to separate and
reinforce individual glass panes in
this Renaissance-inspired home.*

sides approach the subject with near-religious fervor. Proponents of true divided lights stand on historical example and architectural purity. Modernists like the look, but prefer windows that are easier to clean and typically more energy efficient. Each side has its merits. Here are a few points to consider:

In a "true" divided six-over-six pane colonial double-hung window, there are a dozen small double-glazed panes in individual frames. The design is historically accurate, as this was a typical arrangement in windows from a century and more ago. When modern, energy-efficient double glazing is used (an energy code requirement in many areas today), the frames around each pane tend to be thicker and less "elegant" than those originally used, however. There are also twelve separate panes in each window to clean and, when necessary, recaulk and repair.

Where look-alike grilles are used, the homeowner gets the benefit of large, double-glazed panes and a "colonial" appearance, but the major drawback is that these windows rarely have the antique character of true divided lights. In a timberframe home, this can be egregiously noticeable, and can entirely spoil the effect a designer, builder, or homeowner has worked so hard to achieve.

Manufacturers, particularly at the high end of the quality spectrum, have worked hard in recent years to create windows that offer all of the aesthetic and practical benefits of true divided lights, but without their obvious disadvantages. And a number of them have succeeded. As design experts at Andersen Windows point out, "Grille pattern influences perception of style. The same window takes on a completely different character as you change the grilles, and now custom grilles can be made to order in almost any pattern."

WINDOW BASICS

Double-hung windows have their roots in the colonial era, while casement or awning windows (usually controlled with a crank-type operating handle) are more contemporary. The advantage of casement windows is that they are better at capturing even slight breezes and allow for greater airflow when open. The mechanical operator also makes opening and closing more convenient, especially in a difficult-to-reach spot, such as over a kitchen sink. One disadvantage is that casements can be a hazard where they open onto a deck or over areas where people walk or garden.

Sliders or gliding windows, along with sliding patio doors, were developed during the 1940s and 1950s, according to Andersen. Today, those aging aluminum gliders are often replaced by solid vinyl sliding doors, and the "classic" single-pane side-by-side units in many homes are giving way to multi-pane French-style sliding doors or double-hinged French doors.

One innovation that has come along in recent years is the corner window. These curved- or bent-glass units are attractive and can make an extraordinary design statement in the right place, but they are not well suited to timberframe homes—whose load-bearing corner posts are essential to most frame designs—without careful and often expensive engineering early in the planning stage.

Operable skylights or "roof windows" also have grown in popularity over the years. A few decades ago, a home might have had just one strategically placed skylight to brighten an otherwise dreary hallway or room, but today it's not unusual to see entire banks of overhead glass letting the sun shine in. Despite the otherwise traditional heritage of timberframe homes, roof windows fit right

ENERGY EFFICIENCY

The National Fenestration Rating
Council provides accurate information on measuring and comparing
window energy performances. If
the NFRC has rated a window, it
will have a label that notes:

- U-factor, or how well the
  window retains heat inside
  the home.

- Solar heat-gain, or the
  window's ability to resist
  warming from sunlight.

- Visible light transmittance,
  or how much light actually
  passes through the glass.

Significant improvements have
been made since double-glazed
windows became commonplace
in the 1960s and 1970s. First,
the space between the panes is
now typically filled with an inert
gas such as argon to restrict the
flow of heat between the panes.
A second improvement can be
seen in the spacers between
the panes, which were once
made of aluminum that resulted
in an "energy short circuit,"
but are now typically made
with low-conductivity materials.

TIMBERPEG/RICH FRUTCHEY

For special places, such as over the sink counter in a kitchen, skylight window boxes or even a spacious bay window are attractive and functional alternatives to conventional windows.

Low-emissivity glass is another innovation now commonplace and even required by some energy codes. Low-e glass is coated with a microscopically thin metallic layer that is virtually invisible, but which effectively suppresses radiative heat flow.

To achieve maximum energy efficiency, new-home builders or buyers of replacement windows are encouraged to look for Energy Star and the Efficient Window Collaborative labels when shopping for windows.

**CLASSIC POST & BEAM/
JIM BATTLES**

*Vinyl shakes on this home elimi-
nate maintenance, painting, and
staining, yet are indistinguishable
from natural wood shingles.*

into today's modern, eclectic timber designs. In fact, because timber placement can restrict window placement, skylights are often the solution that allows for plenty of natural light without interrupting the frame design. And high ceilings in Timberframes, once a barrier to operable roof windows, are no longer a barrier thanks to electronic openers and automatic sensors that close the overhead windows at the first drops of rain.

Angled bay and curved bow windows, once the focal point of well-appointed living rooms, are still much in demand, although multi-pane bays and "mulled" or ganged casement units have lost favor to vast expanses of fixed glass that let in a greater amount of light with the least amount of visual obstruction. Manufacturers today still offer bay and bow windows made up of individual elements joined in any combination. These windows add a sense of spaciousness to smaller rooms. Depending on the location, energy efficient units are recommended for rooms with northern exposures, which tend to be colder because of the lack of sun, and for windows that face west, which receive the highest heat gain from afternoon sunlight.

EXTERIOR DOORS

A solid oak or mahogany door makes a handsome statement on the exterior of any home, timberframe or otherwise. But solid wood fails to deliver the maintenance efficiency and energy-

saving value of today's manufactured doors. New options include fiberglass and composite doors with paintable or stainable embossed "skins" that perfectly mimic real wood surfaces. Many of these doors offer insulating values that match or exceed the exterior walls into which they are installed.

A growing trend in high-end homes that fits well with the Timberframe way is specialty wood doors. Several manufacturers now offer custom doors to match anyone's wildest home-building imagination. Crafted in exotic woods, designed to look as though they were built for a king's castle or a sea captain's tall ship, antiqued and glazed to suit any era or locale from the ancient to the futuristic, these doors can also be built with insulating layers to meet modern energy standards. Just as gated homes are all the rage in many suburban communities, so too have these exotic doors captured the imagination of homebuilders and buyers alike.

Away from the front entry, sliding doors remain popular, but they also have improved to the point where they bear little resemblance to their progenitors of just a decade or more ago. Manufacturers now offer sliding doors with interior wood frames and brass hardware that are every bit as elegant as genuine French doors, yet far more practical and weathertight. Of course, true French-style doors are available as well. Buyers today also have the

option of French doors that swing out rather than in, which saves interior space and allows for placement of furniture nearer to the doors.

INTERIOR DOORS

Builders are always looking for ways to save on costs, and many builders routinely install plain flush (smooth-skinned) lauan-plywood, hollow-core doors throughout their homes' interiors, while offering higher-quality panel doors as an upgrade. But solid wood panel doors are expensive to buy and finish, so millions of homebuyers over the years have found themselves stuck with the least costly, and least attractive, alternative.

Today there is another, low-cost option that is far superior in looks, price, and performance to hollow-core doors—the Masonite or hardboard-surfaced interior door. These doors have one-piece, pressure-formed exterior skins that exhibit the appearance of true raised-paneled doors, but are built without separate panels. They come pre-primed and accept paints and solid-color stains better than wood doors. They also have an insulated core that reduces sound transmission while providing a heftier "feel" than hollow-core doors. About the only thing these doors can't do better than wood is actually look like natural-surface stained wood, but be assured that manufacturers are working on this.

GARAGE DOORS

When it comes to doors, attention to detail is important. Although they are usually taken for granted, walked through, slammed, and often kicked shut, doors are focal points, inside and outside the home. For far too long, however, one of the most overlooked design areas in the home has been its garage door.

DESIGNER DOORS, INC.

Carriage-style garage doors take center stage on this rustic retreat and are more appropriate to the setting than typical roll-up doors.

DESIGNER DOORS, INC.

Where a garage faces front, custom roll-up doors with handsome wood finishes and divided windows are a stylish alternative to basic doors.

Although nearly every home has them, and although they project an enormous and highly visible surface to the world—they are often the first thing visitors see when approaching a house—manufacturers have barely improved on the looks of tilt-up and roll-up doors over the years. Different shaped windows are offered, as are faux panel styles, and some doors have embossed surfaces to look (unconvincingly) like genuine wood.

All that may finally be changing, though. An entirely new category of garage doors have recently been introduced that not only pass as real, substantial wood doors, but are also so superior in styling that they make older, ordinary garage doors—even heretofore "upgrade" models, look shabby and obsolete. And despite the fact that many of these new doors are styled to look convincingly like side-hinged, carriage-house doors, they are sectioned to roll up just as easily as typical roll-up doors. They can also be fitted with automatic operators, so none of the convenience has been lost. Several manufacturers now offer a line of these doors, and if the price is higher, so is the return in exterior good looks.

capping it off

Timberframe homes generally have roof systems designed and constructed for energy efficiency and durability. The crowning touch, though, is the roofing material applied over these roofs, and any number of options is now available.

Standing-seam metal roofs have long been popular in northern areas for their snow-shedding ability, but their colorful good looks and renowned longevity is making them increasingly attractive to builders in other regions where snow is not a factor. This type of roof is created from metal panels that are usually roll-formed at the job site. As the name implies, the panels are fitted to one another by means of an interlocking vertical seam that forms a watertight joint.

The most popular metal used for standing seam roofs is cold-rolled sheet steel, which is often coated with an aluminum-zinc alloy. Buyers today have almost unlimited options in designer colors, from barn red to soft earth tones and just about every color in-between. These colors are baked on and remain trouble-free for the forty-to fifty-year life of these roofs. Among the metals, natural copper is the highest quality, and generally the most expensive, roofing material. But copper is prized for this use because it can

remain exposed to the elements for centuries. The natural green patina that forms as copper oxidizes helps to shield it from decay, unlike rust oxidation, which consumes most other metals.

Atop the very best residences, slate was once the king of roofing materials. Noted for its durability and an old-world look that projects quality, slate has fallen from favor—except, perhaps, for those who can afford it—because it is extremely difficult and expensive to install, maintain, and repair, and there are few practitioners left to do it.

Because it is a natural material used in its natural state, slate has some interesting quirks that must be addressed. For example, stone slates are recommended for use mainly in the region where they are quarried, and may not perform as well in an area with a different climate. It is unwise to mix limestone (calcareous) slates with sandstone slates. The calcium carbonate in the limestone is slowly dissolved by rainwater, and it can be absorbed by sandstone; as the shingles dry, it recrystalizes, which causes flaking or splitting.

Concrete slate is also available today and it has some advantages, such as lower cost and no materials compatibility issues. It is similar in appearance to real slate, although like most manufactured faux products it has a tendency toward uniformity of appearance and color that does not quite match the real thing.

If natural or artificial slate is used as a roofing material on your next home, be certain that the roof structure is properly designed to support its weight. Timberframes, by their nature, are generally sturdy enough for any type of roof overlay, but the roof sheathing or lath also must be up to the task.

Cedar shingles are another high-quality, long-wearing roofing material that is appropriate in nature to the Timberframe way. Eastern white cedar weathers to a beautiful silver color when exposed to sun and salt air, while thick, eastern red cedar handsplit shingles evoke a forested look as they gather moss season after season. As durable as cedar is, it can be attacked by mold and fungus, but if treated with a preservative it is guaranteed for up to thirty years.

Cedar shakes and shingles stand up well to the elements, and wind-tunnel research shows that properly applied cedar roofing will withstand winds exceeding 130 miles per hour. Wood roofing is also more resilient to hail than other materials. Fire-retardant fire protection can be a priority with consumers who are building, especially in fire-prone regions such as California and Texas. With proper treatment, cedar shakes and shingles fulfill even the toughest building and fire codes.

Fiberglass-reinforced asphalt shingles are the most widely used roofing materials in homes today. Nearly 12.5 billion square feet of various types of asphalt shingle products are manufactured annually—enough to cover more than five million homes every year, according to the Asphalt Roofing Manufacturers Association. These shingles offer quality, durability, and versatility,

and are far less costly than most other roofing materials. Asphalt shingles also offer consumers the broadest array of colors, shapes, and textures available. With a broad range of styles, these shingles can match most types of architectural designs, and affordably achieve virtually any desired aesthetic effect.

BUILDING ON A SCHEDULE

Building a home from start to finish is no haphazard affair. Anywhere from one to two dozen separate subcontractors are typically involved at one point or another. Often, several "subs" will be on the job simultaneously, or one will start and stop his work at different stages to allow other subs to do their part, and then will jump back onto your job to pick up where he left off. It takes careful scheduling—and someone willing and able to enforce it—to keep everything on track and on time.

Material deliveries also have to be carefully coordinated to keep work moving at the proper pace. Certain materials such as stick lumber can be stockpiled at the job site (if theft isn't a problem in your building area), but other goods may be too valuable, too delicate, or too weather-sensitive to leave lying around on a half-completed construction project. Because both workspace and storage space are often limited on building sites, contractors spend considerable time juggling delivery dates to coincide with the ebb and flow of progress, which itself is subject to change without notice.

And it's not just your project's schedule that a contractor has to deal with. If the job doesn't run like clockwork, or the proper materials aren't on hand when needed, the subcontractors' schedules can be thrown off as well—and getting a sub to rework his other job schedules to suit yours can get sticky, especially during peak building seasons when good tradesmen are in demand.

Log and timber pre-cut or kit homes can present their own special materials scheduling challenges. Some years ago I created a cabin-building project for *Field & Stream* magazine, called The Sportsman's Camp. The kit supplied by Northeastern Log Homes was a model of efficiency—everything needed for its construction was delivered on one oversized flatbed tractor-trailer. But finding a place to offload, separate, store, and secure all of the materials, from logs to windows to decking to roofing shingles, could have been a builder's nightmare—had I not been prepared for it. On my own home-building project, which included Insulspan structural insulated panels (SIPs) for the walls and roof, I *was* unprepared—never having built with these unfamiliar materials. A half-dozen pallet trucks

showed up ahead of schedule one evening, looking like a caravan of circus wagons stacked tall with four-by-twenty-foot foam-filled panels. I had to round up a crew to offload them in the dark, and then had to readjust our entire job schedule to deal with the mountain of materials piled around the site.

Another scheduling point that must not be overlooked is inspections, whether from the local municipality or from your bank (if you're building with construction loan funds). Periodic progress inspections are both necessary and unavoidable. If the work isn't completed properly or in a timely manner, you could face a job interruption that throws your schedule—and the entire project—into limbo.

Whether you choose to be your own contractor or just want to stay on top of the construction schedule (as you should—it's *your* home that's being built), you'll need a "roadmap" that helps you understand the big picture and at the same time allows you to keep track of day-to-day activities. Builders typically use construction timelines, or project scheduling flowcharts, to coordinate progress from Day One through C.O. (Certificate of Occupancy).

You may be able to find examples of construction timelines in your local library or bookstore,

or contact the publications department of the National Association of Home Builders (www.builders-books.com) for a list of titles that include this information. Some log and timberframe manufacturers, such as Riverbend Timber Framing, publish their own books and materials that explain the entire kit-building process and include scheduling planners. Many kit manufacturers also offer planning, building, and scheduling seminars for prospective buyers.

CONSTRUCTION TIMELINE

The following timeline gives a general overview of what needs to happen at specific points in the building schedule through a typical six-month home construction project. Keep in mind that this chart is just a model, and is deliberately broad, reflecting average times needed to complete the various tasks:

WEEKS 1 TO 3

- Site clearing: If your site is heavily wooded, you may need a logging crew and additional time to clear and prep for construction. In other cases, the excavator usually handles this work.

- Foundation excavation: Earth-moving equipment clears and shapes the landscape, moves valuable topsoil out of the way for later reuse, and digs the holes for footings and foundation.

WEEKS 2 TO 3

- Septic excavation: If you don't have access to a municipal sewer line, the excavator will prepare the ground for a cesspool or leach field.

- Well drilling: This is done by a subcontractor with specialized equipment. Even a deep (250 feet or more) well can be drilled in a day or two if there are no impediments.

WEEKS 3 TO 5

- Septic construction: Collection boxes, tanks, or a tile field are installed and backfilled.

- Foundation footings: After the builder sites the house, a concrete mason sets forms and pours concrete for the foundation footings.

WEEKS 4 TO 6

- Foundation walls: The concrete mason returns to construct wall forms, and concrete is delivered and poured. If cement blocks are used for the foundation, a block mason begins building the walls atop the footings.

WEEKS 5 TO 7

- Foundation walls: Poured-wall forms are removed, or block-work is completed.

- Foundation waterproofing and drains: The concrete or block walls are sealed with asphalt compound, and drainage lines are placed in gravel around the footings.

- Backfill and rough grading: The excavator returns to push the dirt back against the foundation walls.

WEEKS 6 TO 8

- Deck (floor) framing: A carpenter framing crew builds a deck atop the foundation and prepares to erect the house shell, which may be stick construction, timberframe, or logs.

- Exterior wall framing, log, or timber erection: With the deck complete, the framing crew begins to build the stick-lumber exterior walls and roof, or a specialty crew is brought in to raise the timbers or logs.

- Exterior SIPs: If a timber frame is erected and structural insulated panels are used for exterior walls, another crew may be required to install them as soon as the timbers are up.

WEEKS 7 TO 9

- Window and door installation: The carpenters return to install windows and exterior doors in the rough walls.

- Interior framing: Whatever type of exterior framework is chosen, interior partition walls and floors are required to divide rooms and provide space for plumbing, ductwork, etc.

WEEKS 8 TO 10

- Roofing installation: As soon as the exterior shell is sheathed, roofing work begins, protecting and enclosing the interior. When a house reaches this stage, it is considered "dried in" and interior finish work can start.

- Siding installation: Choice of exterior finished siding is installed.

- Rough plumbing: Plumbers begin installation of drain, waste, and vent (DWV) infrastructure, and house freshwater lines.

- Electric service: Outside power is brought to a pole at the property line. An electrical tap is provided for the building tradesmen, and a meter is installed to monitor use. At this point, the excavator may return to trench for underground power, telephone, and cable lines from the pole to the house.

- Water and sewer service connection: If municipal services are available, the excavator also trenches for the wastewater and freshwater lines, and digs a pit for the curbbox connection.

- Well and septic connection: The house is connected to water source and outflow lines.

- Concrete flatwork: Basement and garage floor are poured and finished.

- Rough electric: House service panel is installed and wiring installation begins.

- Pressure tests: Well pump and plumbing operation are confirmed.

- HVAC systems: Ductwork, baseboard, furnace or boiler, air handlers, and compressors for heating, ventilation, and air conditioning are installed.

- Insulation: If conventionally framed, house walls and attic are insulated with batt, blown-in, or foam insulation (not required with log or SIP construction).

- Drywall: Interior walls and ceilings are "rocked" and taped. Optional skimcoat plaster or conventional plastering begins.

- Exterior painting: Painters begin work on exteriors.

- Gutters and garage doors: Specialty subcontractors install exterior systems.

- Interior tile: Wall and floor tiles are installed by tile mason.

- Fireplace: Specialty subcontractor installs masonry or prefabricated hearth and chimney.

- Interior painting: After drywall taping or plaster is complete, one prime paint coat and two topcoats are applied.

- Cabinetry: Kitchen and bath cabinets are installed.

- Interior trim: Moldings, finish staircases, built-in cabinetry, and other finish carpentry work begins.

- Plumbing: Porcelain fixtures, faucets, and other finish plumbing hardware are installed.

- Final grading: Excavator returns to replace topsoil and finish grading around house.

- Plumbing complete.

- Electrical complete.

- HVAC complete.

- Painting complete.

- Flooring: Hardwood floors installed and finished.

- Exterior flatwork: Concrete, asphalt, or paver driveway and walks are installed.

- Kitchen appliances: Dealers deliver, install, and connect refrigerator, range, and oven.

- Landscaping: Subcontractor installs foundation plantings and prepares lawn.

- Punch list inspection: Builder and homeowner walk through project to ensure that all work is satisfactory and complete.

- Final C.O. inspection: Municipal inspector grants approval for Certificate of Occupancy.

HABITAT POST & BEAM, INC.

Templates may be used to plan interior spaces and furniture placement in concert with a home's overall design.

BUILDERS' SECRETS REVEALED

Larry Fausey and Mike Blair, who have fifty years of timberframe and log-home building experience between them, offer these tips to anyone building a Timberframe, whether it's their first or five-hundredth home.

- When selecting a builder insist on a contractor who will produce well-crafted interior trim.

- Electrically operated skylights with rain sensors are more practical than manual ones and worth the extra cost since they allow you to keep them open without worrying about the weather forecast.

- The interiors of timberframe homes are voluminous, so break up large expanses of walls with wainscoting, wall-covering, or textured paint.

- To hang ceiling fixtures between two beams (purlins), install four-by-eight-inch blocking between the purlins. Place the wide side facing down. This results in a flat, seven-and-a-quarter-inch surface wide enough to accommodate the fixture canopy.

- Budget adequately for lighting fixtures. Though costly, quality fixtures substantially enhance beauty, safety, value, and satisfaction.

CONSUMER CHECKLIST:

Here are typical points every timberframe homeowner should cover when selecting a timberframe or post-and-beam manufacturer.

- What services does the manufacturer offer such as design, engineering, foundation plans, and what does the manufacturer expect the contractor to do on site?

- Is the company a member of the Timber Framers Guild, Timber Frame Business Council, or the Systems Building Council of the NAHB? Does it follow the standards laid out by such building code authorities such as CABO, BOCA, and others?

- What is the company's reputation for on-time delivery and meeting commitments?

- Are the trusses pre-drilled and test-assembled at the factory?

- Will a crane be needed to off-load the truck and position trusses in place, or can they be assembled in place?

- Is the materials list accurate and clear to you? Does the construction manual have intelligible drawings or photos and easy-to-follow instructions?

- Are door openings and windows pre-cut?

- Is there an 800 number to call for technical and other assistance?

- How are posts and beams fastened? Wood trunnels or metal carriage bolts?

• What kind of roof system is included with the package? Are there clear instructions when it comes to making sure the airflow is not compromised between the soffit and ridge vents?

• For cathedral ceilings, a purlin roof system is easier to assemble than rafter roofs that require overhead nailing to cover the rafters with Sheetrock or V-matched boards. Ask the sales representative or drafting department for any unique construction requirements.

• Ask about any installation requirements for well-known brand products that may differ from frame construction.

• Pre-plan wiring. The structural insulated panels or SIPs that enclose most timberframe homes today usually have a wiring chase or channel large enough to accommodate cable, Internet access, and additional phones and electrical outlets that will most likely be needed in the future.

beyond the basic timberframe

taking it outside

The essence of timber framing—sturdy posts, massive beams, and an emphasis on structure as part of the overall aesthetic—can be extended outdoors, too.

Finishing outdoor additions such as pergolas, trellises, or decking in the same wood tones as the interior beams helps to visually connect the exterior with the interior. Another way to "introduce" visitors to your timberframe home is to build outdoor structures that hint at what they will find when they step inside, such as open trusswork in a covered entry or breezeway, or a post-and-beam arbor at the entrance to a garden.

Pergolas are generally defined as freestanding garden structures made up of vertical posts and horizontal "joists." They are large enough for several people to gather under for use as socializing or dining areas. The joists are open to the sky and help to diffuse direct sunlight, making a pergola comfortable throughout the day.

porches and decks

Decks are great for parties, but they can have an aircraft carrier feeling to them, unless vertical elements are introduced to break up the flattop. One way is by designing a multi-level. Each level has a use—dining, lounging, stand-up party area, grille, and so forth. Similarly, wide steps leading to a stone patio serves to break up the horizontal expanse with levels. Incorporating vertical structures into the deck also helps to make the space more user-friendly. Trellises with climbing vines on one side of the deck provide privacy, and tie the deck into the house. A railing, flower boxes, or teak benches all function in a similar way, while reinforcing the emphasis on wood and structure as part of the overall aesthetic standard.

When building a deck or porch, which is technically a deck with a roof, here are a few points to consider:

- Go extra wide, eight feet is about the minimum, but a ten- or twelve-foot-wide porch or deck will feel more comfortable when used for dining.

- The railing creates a visual definition between the indoors and outdoors, which may or may not be the look you want.

- Make the ground level around the deck inviting with a patio and add an interesting focal point, such as sculpture, benches, birdbath, or the like.

- Called a Florida room, three-season porch, sunroom, or solarium depending on the region, a porch that can be completely closed to the weather offers much of the ambiance of an open porch but offers added space for overnight guests. For this purpose, double-hung windows work well for venting top and bottom and establish a more tailored look.

- Old-fashioned wood combination screen and storm doors can be installed side by side to close in a porch. Glass or screen panels can also be installed or removed depending on the season.

- Whether you build a porch or a deck, consider some of the newer composite flooring materials now available that never have to be stained, will never splinter, and leave extra time for fun. They cost about three times more than a wood deck.

sheds

Prefab steel or wooden garden sheds are readily available, relatively inexpensive, and generally quick and easy to erect. They look that way, however, and most are only suitable for that back corner of the property that's not visible to neighbors or guests. Whether you live in a timberframe home or not, if you need a garden shed for tools and equipment, one with a timberframe structure is far more satisfying to build and to own.

TIMBERFRAME GARDEN SHEDS

Designed by the author with architect Ira Grandberg of Mount Kisco, New York, for *Today's Homeowner* magazine, and currently featured on the *This Old House* Web site, Michael's Timberframe Garden Shed is nicely proportioned and roomy enough to park a garden tractor and other outdoor equipment inside. It can also serve as a potting shed and store all of those yard implements that otherwise take up valuable garage space.

An advantage of this frame plan is that it uses six-by-six-inch landscape ties, which are readily available from most lumberyards and garden centers. It can be built by just about anyone with little effort and without difficult carpentry techniques. If you want to really understand how timber framing is done, this is a great place to begin.

The floor plan is ten by fourteen feet (140 square feet), with an interior wall height of eight feet and an exterior height of twelve feet, six inches at the roof peak. To meet local requirements for temporary—that is, non-taxable—outdoor structures, this shed is built atop a foundation of six-by-six-inch landscape ties stacked

SMITH BAER

Common pressure-treated landscape ties were used to fashion this simple yet elegantly useful Timberframe Garden Shed. Plans are available from the builder.

three deep and—bedded in a gravel base. For a more permanent structure, the shed will work with any type of foundation.

The entire framework of the shed is built with common pine pressure-treated lumber, mainly six-by-six-inch-square timbers for the wall posts and beams, and four by fours for the roof purlins. Standard eight-foot-long ties can be used for all sections, although we special-ordered ten-foot and fourteen-foot ties that made it easier to construct the bottom and top wall plates. There are only eight wall-height vertical posts and ten full-length horizontal beams—plus shorter pieces used for blocking between them.

Most of the six by sixes are square-cut and connected with simple interlocking joints that can be made with a circular saw and finished off with a hammer and chisel in more-or-less traditional Timberframe fashion. The only angle cuts in the plan are forty-degree miters where the rafters meet the ridge, and fifty-degree miter cuts for the cross-frame collar ties.

The shed can be closed in with a variety of materials, such as the one-by-six-inch, tongue-and-groove pine boards, or plywood and cedar shakes, or clapboards. For plenty of interior light include inexpensive barn sash windows, which are also available at most lumberyards and home centers.

Windows can be hinged or removable, which allows for the addition of screens in summer.

The two barn-type doors can be made of the same pine boards used for the wall panels. Although they are simple to build, doors of this size can be difficult to hinge and hang properly, so we used a barn-style, galvanized sliding-door track (available from Lawrence Brothers Inc., 800-435-9568). This heavy-duty hardware installs easily and provides a large opening for moving the garden tractor in and out. Four easy-access bolts allow for quick door height and angle adjustment.

A brick-in-sand floor inside maintains the overall theme of simplicity and natural, organic beauty, but it has a practical aspect as well. It's not only one of the easiest permanent floors to install, it also provides a solid base underfoot for a tractor or other heavy outdoor equipment. And when mud and garden debris build up inside, it can be washed down with a garden hose. Extend the brickwork outdoors around the shed to help keep most of the mud outside.

The estimated materials cost to complete the Timberframe Garden Shed is around $2,500. Actual costs will depend on local lumber prices and choice of materials for the foundation, floor, wall panels, and roof. See the Readers' Resource Guide for how to order plans.

timberframe barns

Many timberframe homeowners trace their love of these structures to an impressive barn. Perhaps they peered inside a big red barn with a tall silo while visiting a dairy or apple orchard, or maybe they fondly recall visits to a grandparent's farm and the fun they had playing in the loft. No matter, barns hold a special place in our hearts. It seems quite natural that we want our homes to hearken back to a time when the barn was central to everyday life in America—from meeting our need to protect our food supply to being cleaned out and decorated for a Saturday night "hoedown."

After building a timberframe home, owning a barn can be a passion that just won't die. And, if one needs practical reasons for owning a barn, think of all the vehicles and toys, such as snowmobiles and boats, that a barn will keep safely out of the weather. And why not get a horse or two while you're at it?

If only a timberframe barn will do, there are many approaches. Several artisans offer new barns that are either custom designed or based on standard. Another option is to locate a barn

in need of rescue. Through neglect, demise of the family farm, paths of progress, or bad luck, many American barns must be razed. Fortunately, there exists a cadre of timber framers who specialize in locating, disassembling, and restoring these magnificent structures for new lives as shops, studios, homes, and, of course, barns.

NEW BARNS

For a barn without the wait, a pre-designed, standard kit can be the logical solution. Classic Post & Beam of York, Maine, for example, has introduced its American Barn Series of freestanding structures ideal for use as traditional barns or studios. Kits for a 1,300-square-foot barn start at about $36,000.

As with many timberframe home manufacturers, these kits offer the convenience of having all the components cut to size and delivered at one time, with prices including components one would normally have to purchase locally—such as pre-cut, exposed-beam spruce joists, loft flooring and stairs, roofing materials, windows, doors, and hardware.

Another route to owning a new full-size barn is to contact a timber framer such as Houses and Barns by John Libby in Freeport, Maine. Libby, whose firm has two architects on staff,

LEFT:
DAVIS FRAME COMPANY

*Both new and reconstructed
barn-style structures have great
appeal as homes because they
evoke a rural past while providing
enormous volumes of usable
interior space.*

**NORTHFORD TIMBERFRAMERS/
DICK PIROZZOLO**

*The interior of this salvaged and
rebuilt barn, now part of a home,
looks much as it did when it was
used to store farming equipment.*

will custom design a barn and erect the frame, while leaving closing-in to others to whom he provides detailed construction drawings.

During his thirty years in the business, this Maine craftsman has produced barns and built houses, as well as boathouses and swimming pool barns. His company also took part in building a 38,000-square-foot public market in Portland, Maine.

DICK VISITS GEORGE'S BARN-BUILDING SHOP

A visit to George Senerchia's shop is a like entering a woodworker's heaven. The shop—a restored seventeenth-century barn, of course—is filled with tools ranging from modern power and mortising machines to museum-quality antique framing chisels. He has a collection of early woodworking benches and, pointing out a detail, he says, "See that hole. This bench was built by a framer who didn't own an auger, so he chiseled the holes to accept the bolts, and...the bolts were made by a blacksmith who filed the threads by hand." George's own bench is a replica made of tiger maple.

Although George uses power tools—as long as they do not compromise the outcome—he draws the line when making new trunnels, the wood pegs used to join posts and beams. In this case, his methods are identical to eighteenth-century rural American timber framers. He first shaves square oak pegs into octagons with an adz. Then he pounds the pegs through a sharpened iron ring to round the trunnels off in facets.

"Why not throw the pieces on a lathe—or buy dowels and cut them to size like sausage links?" I ask.

"I've restored fourteen barns and I've met fourteen craftsmen through telepathy. When I pull out a trunnel from the frame of a 200-year-old barn, it's like shaking hands with the builder. This is just my way of showing him respect."

At the time, George's wife, Susan, was mortising a beam. He called it a "tuition project" that is to pay for his daughter's college expenses for the year. Susan let me complete the mortise she had begun—under George's supervision. First, George demonstrated the process, explaining the reason for each step along the way, "Score the joint to prevent any funny splits. Now hold the chisels with the angle *toward* the part you want to remove. Let the tool *push* the wood away."

*A large, custom-built cupola can add
a crowning touch to any home.*

Using a bronze mallet, I began to
shave off pieces, tentatively at
first, then more boldly. Then I was
into *whack, whack whack!* George
inspected. It was the moment
of truth. I tested the fit of the
original storm brace with the
new mortise. The tenon wobbled
around in the mortise like a rocking
chair in a bathtub. It looked like it
was three inches too big—a major
screw-up.

"George what do I do? I cut too
much, it's ruined."

"Nope. You just didn't cut
enough." Then—*wham*, in one
stroke George shaved a half-inch
chunk of wood from one of the
inside angles. The brace slid
in. "Perfect. See how it butts up
against the top? That's where
it counts. That's where the
support is needed."

the timberframe way of the future

American barn guru George Senerchia builds new barns as well as restoring old ones. Among his new barns, George is particularly proud of a cherry frame barn that he sold to a neighbor. He topped it off with one of his unique handmade cupolas, which are built to the standards of fine furniture.

His restoration projects include barns from the seventeenth, eighteenth, nineteenth, and early twentieth centuries. His structures have become a clock shop whose massive iron barn door hinges look like the hands of a clock, an artist's studio, and a shop for an artisan who makes platinum prints for famous photographers, including Annie Liebowitz. He even built three restored barns on his own property just because he couldn't part with them.

During one of my visits, neighbor Robert Newton, Jr., the eighty-three-year-old owner of a family farm up the street, stopped by. When Newton no longer needed his barn—the one he erected with his father in 1923—he entrusted it to George.

It has now been fully restored and is used by a rare book mail-order business.

One night, when George returned to his shop after working on Newton's barn, he stumbled upon a carton filled with old chisels. Attached was a simple note from Bob Newton: "These were my father's. Maybe you can use them." The very tools he'd built the original barn with three-quarters of a century ago! Senerchia restored the chisels—one has a four-inch blade and is nearly three feet long—and used them to make the replacement beams for the Newton barn.

Perhaps more than anything, this is the Zen of timber framing—full-circle, conservation of matter. It lets us know the world will keep spinning, the sun will come up another day, and craftsmanship will live on in the hearts and hands of timber framers today and . . . those who will be inspired by the muse of this craft in generations to come.

This *is* indeed The Timberframe Way!

reader's resource guide

advice &
information

**The American Society of
Landscape Architects**
4401 Connecticut Avenue NW,
Washington, DC 20008
Telephone: (202) 686-2752
Web site: www.asla.org

*Contact the society for details, local contacts,
and literature.*

ASSOCIATIONS

The post-and-beam and timberframe industry
sponsors two Web sites:

National Arborist Association, Inc.
P.O. Box 1094, Amherst, NH 03031
Telephone: (800) 733-2622

*Supplies the names of local members who can
guide the new homebuilder and owner.*

The Timber Frame Business Council

217 Main Street, Hamilton, MT 59840
Telephone: (888) 560-9251
Web site: www.timberframe.org
E-mail: Nancy Wilkins, Executive Director at
nancy@timberframe.org

*A nonprofit organization whose goal is to increase
awareness of the benefits of timberframe construction. Member companies include builders and
designers of timberframe homes, timber suppliers,
structural insulated panel (SIP) manufacturers,
and other suppliers.*

The GE Lighting Institute at Nela Park
1975 Noble Road, Cleveland, OH 44112-6300
Telephone: (800) 255-1200
Web site: www.gelighting.com/na/institute

*Dedicated in 1946 with Mrs. Thomas Edison
participating, the institute traces its roots to 1879
when Edison invented the carbon filament lamp.
In 1892, the General Electric Company was formed
by merging the Edison Electric Company and the
Thomson-Houston Company. The GE–sponsored
institute is a fount of lighting knowledge for both
professionals and consumers.*

The Timberframers Guild

P.O. Box 60, Becket, MA 01223
Telephone: (888) 453-0879
Web site: www.tfguild.org
E-mail: Joel McCarty at joel@tfguild.org

*Geared toward member education, the guild is
helpful to those who want extensive knowledge of
the timberframe process or wish to try their hand
at timber framing before starting their home
project. This is an excellent source for seminars
and workshops by timberframe artisans.*

Sea Gull Lighting
301 West Washington Street, Riverside, NJ 08075
Telephone: (856) 764-0500
Web site: www.seagulllighting.com

*Publishes catalogs that contain a wealth of
information on equipment and technology.*

ROOFING & SIDING

Asphalt Roofing Manufacturers Association
4041 Powder Mill Road, Suite 404,
Calverton, MD 20705
Telephone: (301) 348-2002
Web site: www.asphaltroofing.org

Offers free booklets on a wide range of roofing topics, including "A Homeowner's Guide to Quality Roofing," which covers new roofs and color selection.

Cedar Shake and Shingle Bureau
P.O. Box 1178, Sumas, WA 98295
Telephone: (604) 820-7700
Web site: www.cedarbureau.org

Offers technical, maintenance, and installation information available as PDF files.

The Metal Roofing Alliance
East 4142 Highway 302, Belfair, WA 98528
Telephone: (360) 275-6164
Web site: www.metalroof.org
E-mail: info@metalroofing.com
Contact: Tom Black, Executive Director

Founded in 1998, the MRA is a coalition of metal roofing manufacturers, paint suppliers and coaters, dealers, metal industry associations, and roofing contractors.

KITCHENS & BATHS

National Kitchen & Bath Association
687 Willow Grove Street,
Hackettstown, NJ 07840
Telephone: 877-NKBA-PRO (877-652-2776)
Web site: www.nkba.org

The NKBA offers a free "Kitchen & Bath Workbook" with helpful hints and ideas, referrals to design professionals who are NKBA members, and information on the latest kitchen and bath trends.

STRUCTURAL INSULATING PANELS

Structural Insulating Panel Association
P.O. Box 1699, Gig Harbor, WA 98335
Telephone: (253) 858-7472
Web site: www/sips.org
E-mail: Bill Wachtler at staff@sips.org

An excellent informational resource and guide to manufacturers.

FIREPLACES & OUTDOOR LIVING

Hearth, Patio & Barbecue Association
1601 North Kent Street, Suite 1001,
Arlington, VA 22209
Telephone: (703) 522-0086
Web site: www.hpba.org

Formerly known as the Hearth Products Association, the HPBA offers a wide range of information on the selection, and installation of fireplaces and barbecue products, safety guidelines, and a dealer locator service.

HOME SHOWS

Country's Best Log Homes and Timber Frame Show
11305 Sunset Hills Road, Reston, VA 20190
Telephone: (703) 327-8781
Web site: www.countrysbestloghomesmag.com
E-mail: Laurie Sloan at
laurie@sloanpublishingservices.com

**Great Midwest Log Home and
Timber Frame Show**
3488 Cty J, Pine Bluff, WI 53528
Telephone: (608) 798-3222
Web site:
www.midwestloghomeandtimbershow
E-mail:
griswold@midwestloghomeandtimbershow.com
Contacts: Greg Griswold or Laura Wierzbicki

*Showcases rustic log furniture, lights, home
accessories, copper and iron art, antler lighting,
carved fireplace mantels, and many other
unique home ideas.*

**Log and Timber Frame
Consumer-Trade Exposition**
4125 Lafayette Center Drive, Suite 100
Chantilly, VA 20151
Telephone: (703) 222-9411
Web site: www.timberframehomes.com

*In addition to home shows, timberframe maga-
zines list regional open houses and courses. Many
manufacturers welcome visitors to their facilities
individually or as part of a tour or host workshops
where consumers can see the process in action
and try their hand at the timber framer's craft.*

Log Home and Timber Frame Expo
10-3435 Westsyde Road, Kamloops, BC V2B 7H1
Telephone: (888) 564-3976
Web site: www.logexpo.com
E-mail: Debbie Fraser at info@logexpo.com

*A log and timberframe consumer/trade exposition
organized by Log Homes Illustrated.*

INTERIOR DESIGNERS AND ART
CONSULTANTS
Barbara Bent Hamilton Interiors
94 Woodside Avenue, Wellesley, MA 02482
Telephone: (781) 237-2866
Contact: Barbara B. Hamilton

Clark Planning & Design
470 School Street, Rumney, NH 03266
Telephone: (603) 786-3635
Contact: Mary Clark

Decorating Den
3622 Locust Avenue, Louisville, KY 40299
Telephone: (502) 228-9796 or (502) 491-2988
Contact: Linda Corzine

J/Brice Design International
326 A Street, Suite 18, Boston, MA 02210
Telephone: (617) 695-9456
Web site: www.jbricedesign.com
E-mail: Jeff Ornstein at Jeff@jbricedesign.com

Renée Fotouhi Fine Art, Ltd.
315 Church Street, New York, NY 10013
Telephone: (212) 431-1304
Web site: www.fineartcandy.com
Contact: Renée Fotouhi

ARCHITECTS
Cargill/Blake
396 Route 49, Campton, NH 03223
Telephone: (603) 726-3939
Contact: Teresa Cargill

*Architects and builders of timberframe and
other unique homes.*

Earth & Sky Architecture
869 Santa Fe Drive, Denver, CO 80204-4344
Telephone: (720) 956-1643
Web site: www.earthskyarchitecture.com
E-mail: Paul R. Adams at
padams@earthskyarchitecture.com

*Environmentally sound building practices
combined with an understanding of what
people like to call home.*

Flannel Shirt Architects
85 Harvest Lane, Shelburne, VT 05482
Telephone: (888) 700-5019 or (802) 985-9815
E-mail: Stephen Moore at
mooreaia@together.net

*Practice limited to timberframe design.
Member AIA.*

MacPhail Architectural Collaborative
109 Trapelo Road, Belmont, MA
Telephone: (617) 489-8535
Web site: www.marc-arc.com
Contact: Michael MacPhail

*Residential timberframe buildings and additions
in Massachusetts, the Cape, and Islands.*

Mountain Timber Design
616 Cressman Court, Golden, CO 80403
Telephone: (303) 278-8986
Web site: www.mountaintimberdesign.com
E-mail: Judd Dickey at mtdarch@aol.com

Architecture and consulting.

Paradigm Builders
5 Crowley Terrace, Hanover, NH 03755
Telephone: (603) 643-2002
Web site: www.timbercad.com
E-mail: Ed Levin at elevin@valley.net

*Timberframe design services to owners,
contractors, architects, and engineers, specializing
in curved framing and compound joinery.*

Rex Hohlbein Architects
18116 101st Avenue NE, Bothell, WA 98011
Telephone: (425) 487-3655

Richard Berg Architects, PC
727 Taylor Street, Port Townsend, WA 98368
Telephone: (360) 379-8090
E-mail: Richard Berg at rberg@olympus.net

*Full service architectural design of homes
and commercial buildings, using timberframe,
conventional, and SIP construction.*

Sam Marts Architects & Planners
2014 West Wabansia, Chicago, IL 60647
Telephone: (773) 862-0123
Web site: www.timbersmart.com
E-mail: info@timbersmart.com
Contact: Sam Marts

*Affordable design services for custom
timberframe homes.*

Springpoint, Inc.
23 Comstock Road, Alstead, NH 03602
Telephone: (603) 835-2433
Web site: www.springpointdesign.com
E-mail: springpt@sover.net
Contact: Andrea Warchaizer

*Architectural design specializing in site-sensitive,
resource-efficient, life-enhancing homes.*

Stark Ireland Architects Inc.
2560 Matheson Boulevard East, Suite 317,
Mississauga, Ontario L4W 4Y9, Canada
Telephone: 905-629-2500
Contact: John Stark, President at
stark@starland.on.ca

Timber Frame Design Associates
23 Comstock Road, Alstead, NH 03602
Telephone: (435) 408-1591
E-mail: tfda@bigfoot.com
Contact: Andrea Warchaizer

*Timberframe design team offers architectural and
structural design, from schematics to complete
construction documents, and specifications.*

publications

BOOKS

Adirondack Style by Ann S. O'Leary (New York:
Clarkson Potter, 1998)

*Build a Classic Timber-Framed House:
Planning and Design, Traditional Materials,
Affordable Methods* by Jack A. Sobon (Pownal,
VT: Storey Books, 1994)

Building With Nantucket In Mind by
Christopher Lang and Kate Stout (Nantucket
Historic District Commission, 1995). Available at
Mitchell's Book Corner, 54 Main Street,
Nantucket, MA 02554, Telephone: (508) 228-1080.

Brunschwig and Fils Style by Murray Douglas
and Chippy Irvine (Boston: Bulfinch Press, 1995)

*Divine Design: Decorating Den's 25th
Anniversary Collection* by Carol Donayre Bugg,
(Decorating Den Interiors, 1994)

*The Domain Book of Intuitive Home Design:
How to Decorate Using Your Personality*
by Judy George and Todd Lyon (New York:
Clarkson Potter, 1998)

*Flowers Underfoot: Indian Carpets of the
Mughal Era* by Daniel S. Walker (New York:
Metropolitan Museum of Art, 1997)

*Oriental Rugs of the Silk Route Culture
Process and Selection* by John B. Gregorian (New
York: Rizzoli International Publications, Inc. 2000)

Rustic Style by Ralph Kylloe (New York: Harry N.
Abrams, 1998)

Timberframe Interiors by Dick Pirozzolo and
Linda Corzine (Layton, UT: Gibbs Smith,
Publisher, 2000)

The Timberframe Plan Book by Michael Morris
and Dick Pirozzolo (Layton, UT: Gibbs Smith,
Publisher, 2000)

PLANS

Timberframe Garden Shed by Michael Morris
and Ira Granberg

Complete, easy-to-follow plans, materials list,
and cutting diagrams are available from:
Plans Department, Mercurial Editorial
P.O. Box 94, Maryknoll, NY 10545
Telephone: (914) 762-8417

*Country's Best Luxury Log and
Timberframe Homes*
Sovereign Media
30 West Third Street, Williamsport, PA 17701
Telephone: (570) 322-7848

Timber Frame Homes Magazine
Home Buyer Publications
4200-T Lafayette Center Drive,
Chantilly, VA 20151
Telephone: (703) 222-9411

Timber Homes Illustrated Magazine
GCR Publishing
1700 Broadway, 34th Floor, New York, NY 10019
Telephone: (212) 541-7100

barn wrights & barn restorers

BARN BUILDING, RESTORATION & SHEDS

While timberframe and post and beam manufacturers also offer barns, the U.S. is dotted with timber framers experienced in new and old barns—including restorations—either in place or dismantled and built on your property. Many of these artisans also build homes—from cutting and erecting the frames to complete move-in projects.

Roger Audette
R.R. 2, Box 5855A, Winthrop, ME 04364
Telephone: (207) 395-2631

Barn Wright, Inc.
244 Scribner Mill Road, Harrison, ME 04040
Telephone: (207) 583-4289
Contact: Scott Hatch

Jeremie Berube
398 River Road, Arundel, ME
Telephone: (207) 985-1927

Dick Brown
Blue Hill, ME 04614
Telephone: (207) 367-5993

Connolly & Company—Timber Frame Homes and Barns
Route 1, Edgecomb, ME 04556
Telephone: (207) 882-4224

M. A. Frigon
P.O. Box 245, Jackman, ME 04945-0307
Telephone: (207) 668-7772

J. F. Heaney & Company
P.O. Box 1275, Waldoboro, ME 04572
Telephone: (207) 832-4691
E-mail: jfh@midcoast.com

Houses & Barns by John Libby
17 Post Road, P.O. Box 258,
Freeport, ME 04032
Telephone: (207) 865-4169

Michael Landry
17 Winterhaven Drive, Orono, ME 04473
Telephone: (207) 866-0118
E-mail: mldesign@mint.net

Northford Timberframers
254 Old Post Road, Northford, CT 06472
Telephone: (203) 484-2129
Web site:
www.geocities.com/northfordtimberframers/
Contact: George Senerchia, artisan

home products, appliances, & materials

Acorn Manufacturing Company
P.O. Box 31, 457 School Street,
Mansfield, MA 02048
Telephone: (800) 835-0121
Web site: www.acornmfg.com

Blacksmith door and cabinet hardware.

Aga Ranges LLC
110 Woodcrest Road, Cherry Hill, NJ 08003
Telephone: (800) 633-9200 (U.S.) or
(800) 663-8686 (Canada)

These traditional cast-iron "English cookers," which have changed little since their introduction in 1929, are growing in popularity among American homeowners.

American Standard, Inc.
P.O. Box 6820, One Centennial Avenue,
Piscataway, NJ 08855
Telephone: (800) 524-9797
Web site: www.americanstandard-us.com

Blacksmith finish faucets.

American Traders
627 Barton Road, Greenfield, MA 01301
Telephone: (888) 723-3779
Telephone: www.woodencanoe.com

Rustic cedar furnishings and canoes.

Architectural Ironworks
2811 Industrial Road, Santa Fe, NM 87505
Telephone: (505) 438-1864

Hand-wrought door hardware and accessories.

Arte de Mexico
1000 Chestnut Street, Burbank, CA 91506
Telephone: (818) 753-4559
Web site: www.artedemexico.com

Hand-forged wrought iron and hand-carved wood lighting and furniture.

Avalanche Ranch Light Company
P.O. Box 31397, Bellingham, WA 98228
Telephone: (360) 752-1610 or (888) 841-1810
Web site: www.avalight.com

Hand-crafted lighting in Northwest and outdoors motifs.

Back Country Antler Works
P.O. Box 400, Interstate 45 North,
Palmer, TX 75152
Telephone: (972) 845-3526

Antler chandeliers, lighting, and tables.

The Bear Factory
37516 Highway 58, Pleasant Hill, OR 97455
Telephone: (541) 746-4842

Custom wood carvings.

Bouvet USA
540 DeHaro Street, San Francisco, CA 94107
Telephone: (415) 864-0273

*Forged iron and brass hardware;
fire retardant treatment.*

Buckley Rumford Company
1035 Monroe Street, Port Townsend, WA 98368
Telephone: (360) 385-9974
Web site: www.rumford.com
Contact: Jim Buckley

Rumford fireplace design and information.

CHEMCO, Inc.
4191 Grandview Road, P.O. Box 879,
Ferndale, WA 98248
Telephone: (360) 366-3500
Web site: www.chemcoproducts.com

*Fire-retardant wood siding and roofing treatment
for exterior wood surfaces, ideal for those building in
urban/wildland interface zones and forested areas.*

Cherry Tree Designs
320 Pronghorn Trail, Bozeman, MT 59718
Telephone: (800) 634-3268

Custom screens, doors, and lighting.

Colorado Log & Antler
P.O. Box 2846, Frisco, CO 80443
Telephone: (970) 453-7011

Rustic home furnishings.

Designer Doors, Inc.
Boston Design Center, One Design Center Place,
Boston, MA 02210
Telephone: (800) 550-1441
Web site: www.designerdoors.com

Custom garage doors.

Heatilator
1915 West Saunders, Mt. Pleasant, IA 52641
Telephone: (800) 843-2848
Web site: www.heatilator.com

*A top manufacturer of wood-burning and
gas fireplaces.*

Hutton Metalcrafts
P.O. Box 418, Route 940, Pocono Pines, PA 18350
Telephone: (570) 646-7778 or (888) 479-1748
Website: www.poconomts.com/hutton/
E-mail: hutlamp@epix.net

*More than thirty styles of rustic lanterns in solid
copper and solid brass.*

James and Company
P.O. Box 31023, Flagstaff, AZ 86003
Telephone: (256) 997-0703
Web site: www.jamesandcompany.com

Antique beams, columns, and millwork.

J.R. Coppersmythe
80 Stiles Road, Boylston, MA 01505
Telephone: (508) 869-2769

Handcrafted lighting fixtures.

The Kennebec Company
1 Front Street, Bath, ME 04530
Telephone: (207) 443-2131
Contact: David Leonard

Custom kitchen design and cabinetry.

Left By Nature
6629 Rockridge Trail, Aubrey, TX 76227
Telephone: (940) 440-3434
Web site: www.leftbynature.com

Natural wood lighting fixtures.

M. Star Antler Designs
P.O. Box 3093, Lake Isabella, CA 93240
Telephone: (619) 795-9577

Antler lighting to complement your home.

The Murus Company
P.O. Box 220, Mansfield, PA 16933
Telephone: (570) 549-2100
Web site: www.murus.com
E-mail: murus@epix.net

Manufacturer of structural insulating panels.

Nantucket Rose & Vine
58 Oxford Drive, Cotuit, MA 02635
Telephone: (888) 627-ROSE
Web site: www.nantucketrose.com

Outdoor trellises and arbors.

Osmose
P.O. Box O, Griffin, GA 30224
Telephone: (800) 241-0240

Wood stains and preservatives.

Pinecrest
2118 Blaisdell Avenue, Minneapolis, MN 55404
Telephone: (800) 443-5357

Hand-carved mantels and wood shutters.

Poker's Triple Seven
P.O. Box 73231, Houston, TX 77273
Telephone: (800) 875-2706

Blacksmith and Western home furnishings.

Precision Pine, Inc.
8919 Valgro Road, Knoxville, TN 37920
Telephone: (877) 885-8902
Web site: www.spiralstaircase.com

Custom staircases.

Ragged Mountain Chandeliers
P.O. Box 1164, Hamilton, MT 59840
Telephone: (406) 961-2400
Web site: www.antlerchandelier.com

Antler chandeliers.

Real Goods Catalog
555 Leslie Street, Ukiah, CA 95482
Telephone: (800) 762-7325
Web site: www.realgoods.com

Energy-efficient home products.

Southwest Door Company
9280 Old Vail Road, Tucson, AZ 85747
Telephone: (520) 574-7374

Custom doors, windows, cabinets, and hardware.

Timber & Stone Restorations
5431 East U.S. Highway 290,
Fredericksburg, TX 78624
Telephone: (830) 997-2280

Antique beams and timber frames.

Tree House Creations
7249 Lamar Alexander Parkway,
Townsend, TN 37682
Telephone: (865) 448-8733
Web site: www.thclogfurniture.com

Log furniture.

Trellis Structures
60 River Street, Beverly, MA 01915
Telephone: (888) 285-4624
Web site: www.trellisstructures.com

*Western red cedar and mahogany arbors,
trellises, and pergolas.*

Vermont Sleigh Company
8341 Gleason Road, Rutland, VT 05701
Telephone: (877) 887-5344
Web site: www.vermontsleigh.com

New England furniture and accessories.

Wild West Log Furniture
1511 Nettleton Gulch Road,
Coeur d'Alene, ID 83814
Telephone: (208) 667-1754
Web site: www.wildwestlogfurniture.com

Rustic log furnishings.

Wood Carvings Unlimited
P.O. Box 1332, Kimberling City, MO 65686
Telephone: (417) 739-5613

Handcrafted mantels, doors, tables, and lamps.

timberframe & post-and-beam home manufacturers

*(Selected from the Timberframe Business
Council membership and other sources)*

Architectural Timberworks
R.R. 1, Box 349A, Dallas, PA 18612
Telephone: (570) 639-2353
Web site: www.archtimb.com
E-mail: info@archtimb.com
Contact: Mitch Rowland

Big Timberworks, Inc.
Box 368, Gallatin Gateway, MT 59730
Telephone: (406) 763-4639
Web site: www.bigtimberworks.com
E-mail: Merle Adams at
merle@bigtimberworks.com

Bitterroot Timber Frames
567 Three Mile Creek Road,
Stevensville, MT 59870
Telephone: (406) 777-5546 or (866) 218-5546
Web site: www.bitterroottimberframes.com
E-mail: Brett Mauri at
bitterroot@bitterroottimberframes.com

BL Corporation, Ltd.
4-9-5 ISO5F, Roppngi Minato-ku,
Tokyo, 106-0032
Telephone: (03) 5 414-3353
Web site: www.timber-frame.net
E-mail: zon@a-9.com
Contact: Tamotsu Yamane

Border Oak Design & Construction Limited
Kingsland Sawmills, Kingsland, Leominster,
Hereford, UK, HR69SF
Telephone: 011-44-1568-708752
Web site: www.borderoak.com
Contact: John Greene

Blue Ridge TimberWrights
P.O. Box 30, Christiansburg, VA 24068
Telephone: (540) 382-1102
Web site: www.blueridgetimberwrights.com
E-mail:
information@blueridgetimberwrights.com
Contact: Sandy Bennett

Brewster Timber Frame Company
701 Pecan Drive, Bellvue, CO 80512
Telephone: (970) 493-7682
E-mail: Steve Rundquist at stevencr@aol.com

British Columbia Timber Frames
P.O. Box 2241, Squamish, BC V0N 3G0
Telephone: (866) 235-2283
Web site: www.bctimberframe.ca
E-mail: info@bctimberframe.ca
Contact: Greg McKee

Cabin Creek Timber Frames
360 North Jones Creek Road, Franklin, NC 28734
Telephone: (828) 369-5899
Web site: www.cabincreektimberframes.com
E-mail: Joseph Bell at jbell@dnet.net

The Cascade Joinery
1401 Sixth Street, Bellingham, WA 98225
Telephone: (360) 527-0119
Web site: www.cascadejoinery.com
E-mail: info@cascadejoinery.com
Contact: Ross Grier

Centennial Timber Frames
1001 Tumble Creek Road, Kalispell, MT 59901
Telephone: (406) 755-8114
Web site: www.centennial.com
E-mail: centennial@centurytel.net
Contact: Mike Koness

Classic Post & Beam
P.O. Box 546, York, ME 03909
Telephone: (800) 872-2326
E-mail: info@classicpostandbeam.com
Contact: James Nadeau

Connolly & Company Timber Frame
10 Atlantic Highway,
Edgecomb, ME 04556-9722
Telephone: (207) 882-4224
Web site: www.connollytimberframes.com
E-mail: connolly@lincoln.midcoast.com
Contact: John Connolly or Scott Clark

Cougar Creek Timber Frames
P.O. Box 867, 100 Davis Lake Ranch Road,
Winthrop, WA 98862
Telephone: (509) 996-3631
Web site: www.cougarcreektf.com
E-mail: Don Carlson at carlson@methow.com

Cowee Mountain Timber Framers
104 Wykle Road, Franklin, NC 28734
Telephone: (828) 369-8186
Website: www.timberframesales.com
E-mail: coweemtn@hotmail.com
Contact: Tom Munger or Steve Smith, Jr.

Craftwright, Inc.
100 Railroad Avenue, #105,
Westminster, MD 21157
Telephone: (410) 876-0999
Web site: www.craftwrighttimberframes.com
E-mail: greyoak1@aol.com
Contact: Glenn Allen James

Davis Frame Company
Route 12A, P.O. Box 1079, Claremont, NH 03743
Telephone: (800) 636-0993
Web site: www.davisframe.com
E-mail: inquiry@davisframe.com
Contact: Jeffery R. Davis

Dreaming Creek Timber Frame Homes
2487 Judes Ferry Road, Powhatan, VA 23139
Telephone: (804) 598-4328
Web site: www.dreamingcreek.com
E-mail: sales@dreamingcreek.com

EarthWood Homes
P.O. Box 807, 111-A2 West Barclay Drive,
Sisters, OR 97759
Telephone: (541) 549-0924
Web site: www.earthwoodhomes.com
E-mail: timberframe@earthwoodhomes.com
Contact: Kris Calvin

EDR, Ltd.
P.O. Box 436, Siren, WI 54872
Telephone: (715) 349-2545
E-mail: Gary A. Pavlicek at pavedr@sirentel.net

Euclid Timber Frames L.C.
Box 969, 3092 West South Highway 189,
Charleston, UT 84032
Telephone: (435) 654-1372
Web site: www.euclidtf.com
Contact: Kurt Apostol

Fall Creek Timber Frame Homes, Inc.
2050 Lake Creek Road, Troy, MT 59935
Telephone: (406) 295-5204
E-mail: fallcrk@libby.org
Contact: Brian Leisz

Fitzgerald's Heavy Timber Construction, Inc.
10801 Powell Road, Thurmont, MD 21788
Telephone: (301) 898-9340
Web site: www.fitzgeraldtimberframes.com
E-mail: Dean Fitzgerald at tmbrfrmr1@aol.com

Gates Timber Frames
Box 4453 B RR 2, Jonestown, PA 17038
Telephone: (717) 865-5700
Contact: Scott Gates

Glenville Timberwrights, Inc.
602 Lake Street, Baraboo, WI 53913
Telephone: (608) 356-9095
Web site: www.glenvilletimberwrights.com
E-mail: Tom Holmes at woodshop@tds.net

Goshen Timber Frames
58 Willow Grove Lane, Franklin, NC 28734
Telephone: (828) 524-8662
Web site: www.timberframemag.com
E-mail: info@goshenframes.com
Contact: Susan Broadhead

Greenside Company
427 Little Pond Road, Sandwich, NH 03227
Telephone: (603) 284-7069
E-mail: philipstrother@msn.com

Habitat Post & Beam
21 Elm Street
South Deerfield, MA 01373
Telephone: (413) 665-4006
Web site: www.postandbeam.com

Hamill Creek Timberwrights, Inc.
Box 151, Meadow Creek, BC V0G 1N0
Telephone: (250) 366-4320
Web site: www.hamillcreek.com
E-mail: Dwight Smith at
dwight@hamillcreek.com

Hampton Timber Frame
P.O. Box 177, Coboconk, ON K0M 1K0
Telephone: (705) 454-3345
Website: www.hamptontimber.com
E-mail: Paul Barber at
pbarber@hamptontimber.com

Harmony Exchange, Inc.
2700 Big Hill Road, Boone, NC 28607
Telephone: (828) 264-2314
Web site: www.harmonyexchange.com
E-mail: Brian Mueller at
mueller@harmonyexchange.com

HearthStone, Inc.
1630 East Highway 25-70, Dandridge, TN 37725
Telephone: (865) 397-9425 or (800) 247-4442
Web site: www.hearthstonehomes.com
E-mail: sales2@hearthstonehomes.com
Contact: Chris Wood

Heartwood Timberframes
2100 East U.S. 223, Adrian, MI 49221
E-mail: bob@heartwoodtimber.com
Telephone: (419) 822-3805 or (888) 676-9870
Web site: www.heartwoodtimber.com
Contact: Bob Sternquist

Highland Timber Frame, Inc.
1019 Thunderstruck Road, Floyd, VA 24091
Telephone: (540) 745-7411
E-mail: Jim Callahan at jrbcalla@swva.net

Homestead Timber Frame
972 Route 106N, Loudon, NH 03307
Telephone: (603) 524-7177
Web site: www.homesteadtimberframe.com
E-mail: Bob Phillips at
bob@homesteadtimberframe.com

Hugh Lofting Timber Framing, Inc.
339 Lamborntown Road, West Grove, PA 19390
Telephone: (610) 444-5382
Web site: www.hughloftingtimberframe.com
E-mail: Hugh Lofting at
hugh@hughloftingtimberframe.com

Joint Effort Timber Framing, Inc.
P.O. Box 223, Cambridge, VT 05444
Telephone: (802) 644-6644
Web site: www.jetf.net
E-mail: Randy Churchill at randy@jetf.net

Legacy Timber Frames, Inc.
691 County Road 70, Stillwater, NY 12170
Telephone: (518) 279-9108
Web site: www.legacytimberframes.com
E-mail: legacytf@aol.com
Contact: Annemarie and Dan Roseberger

Lindal Cedar Homes
4300 South 104th Place, Seattle, WA 98178
Telephone: (888) 454-6325
Web site: www.lindal.com
E-mail: info@lindal.com

Lon Tyler Company
729 50th Avenue, Sweet Home, OR 97386
Telephone: (541) 367-6726
Web site: www.ltyler.com
E-mail: Lon Tyler at lon@ltyler.com

Long Creek Timber Framers
325 Wildwood Drive, Mt. Zion, IL 62549
Telephone: (217) 864-5671
Contact: Glen Vermette

Michigan Timber Frames
7420 Turk Road, Brooklyn, MI 49230
Telephone: (517) 592-2411 or (800) 369-3284
Web site: www.michigantimberframes.com
E-mail: michigantimberframes@earthlink.net
Contact: Chuck Doyle

New Energy Works Timberframers
1180 Commercial Drive, Farmington, NY 14425
Telephone: (585) 924-3860
Web site: www.newenergyworks.com
E-mail: Jonathan Orpin at
jonathan@newenergyworks.com

North Woods Joinery
P.O. Box 180, Cambridge, VT 05444
Telephone: (802) 644-2400
Web site: www.northwoodsjoinery.com
Contact: Pete Kochalka

Northern Lights Timber Framing
3107 Edgewood Avenue South,
St. Louis Park, MN 55426
Telephone: (612) 791-2736
Web site:
www.northernlightstimberframing.com
E-mail: Clark Bremer at
clarkb@northernlightstimberframing.com

Oakbridge Timber Framing
20857 Earnest Road, Howard, OH 43028
Telephone: (740) 599-5711
Contact: Jim Kanagy

Paul Oatman Construction
24500 Robin Hood Drive, Pioneer, CA 95666
Telephone: (209) 295-5100
Contact: Paul Oatman

Perry-Gantt Associates, Ltd.
302 North First, Hamilton, MT 59840
Telephone: (406) 363-0855
E-mail: timber@montana.com

Red Suspenders Timber Frames, Inc.
Route 7, Box 8383, Nacogdoches, TX 75965
Telephone: (936) 564-9465
E-mail: info@redsuspenderstf.com
Contact: Tim Chauvin or Gene McCalmont

Riverbend Timber Framing
P.O. Box 26, Blissfield, MI 49228-0026
Telephone: (517) 486-4355
Web site: www.riverbendtf.com
E-mail: info@riverbendtf.com
Contact: Frank Baker

Riverbend Timber Framing of Georgia
1368 Village Park Drive NE, Atlanta, GA 30319
Telephone: (770) 331-8999
E-mail: hamilton@mindspring.com
Contact: Richard H. Rosenbloom

Rockport Post & Beam
P.O. Box 553, Rockport, ME 04856
Telephone: (207) 236-8562
Contact: Peter Smith

San Juan Timberwrights, Inc.
P.O. Box 1788, Arboles, CO 81121
Telephone: (970) 883-2283
Web site: www.sanjuantimber.com
E-mail: Mike McDowell at sjttlm@frontier.net

Sierra Timberframers
P.O. Box 595, Nevada City, CA 95959
Telephone: (530) 292-9449
Web site: www.sierratimberframers.com
Contacts: Martha and Doug Lingen

Sugar Hollow Timberworks, Inc.
P.O. Box 2123, Fairview, NC 28730
Telephone: (828) 628-4166
Web site: www.sugarhollowtimberworks.com
E-mail: Jeffery M. Hambley at
sugarhollowlogs@aol.com

Sunset Structures, Ltd.
P.O. Box 129, Elkview, WV 25071
Telephone: (304) 965-6831
Web site: www.sunsetstructures.com
E-mail: info@sunsetstructures.com
Contact: Randy Renick

Tennessee Timberframe
2537 Decatur Pike, Athens, TN 37303
Telephone: (800) 251-9218
Web site: www.tntimberframe.com
E-mail: Robert Lambert at
rlambert@tnloghomes.com

Texas Timber Frames, Inc.
5950 Camp Bullis Road, San Antonio, TX 78257
Telephone: (210) 698-6156
Web site: www.texastimberframes.com
E-mail: info@texastimberframes.com
Contact: Bill Farrar

Thistlewood Timber Frame Homes
R.R. 6, Thistlewood Road, Markdale, ON N0C 1H0
Telephone: (800) 567-3280
Web site: www.thistlewoodtimberframe.com
E-mail: homes@thistlewoodtimberframe.com
Contact: Scott Murray

Three Elements Timberworks, Inc.
855 West Dillon Road I-105, Louisville, CO 80027
Telephone: (303) 664-1946
Web site: www.threeelements.com
E-mail: Eric Seelig at eric@threeelements.com

Timber Creations
405 Horn Avenue, Santa Rosa, CA 95407
Telephone: (707) 584-1829
Web site: www.timbercreations.net
E-mail: Leif Calvin at leif@timbercreations.net

Timber Creek Post and Beam, Inc.
P.O. Box 309, Cuttingsville, VT 05738
Telephone: (802) 775-6591
Web site: www.timbercreekinc.com
E-mail: timber@sover.net
Contact: Marty or Dan Pinkowski

TimberCraft Homes, Inc.
85 Martin Road, Port Townsend, WA 98368
Telephone: (360) 385-3051
Web site: www.timbercraft.com
E-mail: info@timbercraft.com
Contact: Kevin Coker

Timberpeg
112 North Main Street, P.O. Box 5474, West
Lebanon, NH 03784
Telephone: (603) 298-8820
Web site: www.timberpeg.com
E-mail: info@timberpeg.com
Contact: Robert Best

Timbersmith, Inc.
4040 Farr Road, Bloomington, IN 47408
Telephone: (812) 336-7424
Web site: www.timbersmith.com
E-mail: timbers@kiva.net
Contact: John Hayes

Timmerhus, Inc.
3000 North 63rd Street, Boulder, CO 80301
Telephone: (303) 449-1336
Web site: www.timmerhus.com
E-mail: Ed Shure at ed@timmerhusinc.com

Trillium Dell Timberworks
1277 Knox Road 1600N, Knoxville, IL 61448
Telephone: (309) 221-9380 or (888) 299-3022
Web site: www.trilliumdell.com
E-mail: Rick Collins at oldblue@grics.net

Unique Log & Timber Works, Inc.
1837 Shuswap Avenue, P.O. Box 730,
Lumby, BC V0E 2G0
Telephone: (250) 547-2400
Web site: www.uniquetimber.com
E-mail: info@uniquetimber.com
Contact: Dave Gardner

Vermont Frames
Box 100, Hinesburg, VT 05461
Telephone: (802) 453-3727
Web site: www.vermontframes.com
Contact: Peter McNaull

Vermont Timber Frames, Inc.
7 Pearl Street, Cambridge, NY 12816
Telephone: (518) 677-8860
Web site: www.vtf.com
E-mail: Jim Gibbons at jgibbons@vtf.com

Vermont Timber Works
P.O. Box 856, Springfield, VT 05156
Telephone: (802) 885-1917
Web site: www.vermonttimberworks.com
E-mail: Kim Hentschel at
kim@vermonttimberworks.com

Woodhouse Post & Beam Homes, Inc.
P.O. Box 219/Route 549, Mansfield, PA 16933
Telephone: (570) 549-6232
Web site: www.woodhouse-pb.com
E-mail: Kathi Van Dusen at
kvandusen@woodhouse-pb.com

Yankee Barn Homes
131 Yankee Barn Road, Grantham, NH 03753
Telephone: (603) 863-4545 or (800) 258-9786
Web site: www.yankeebarnhomes.com
E-mail: info@yankeebarnhomes.com
Contact: Rob Knight

timberframe tools

Bailey's, Inc.
P.O. Box 550, Laytonville, CA 95454
Telephone: (707) 984-6133 or (888) 322-4539
Web site: www.baileys-online.com
E-mail: Bill Bailey at bill@baileys-online.com

Twenty-seven-year-old mail-order company, selling woodsman products and featuring timberframe tools.

Mafell North America, Inc.
1975 Wehrle Drive, Suite 120,
Williamsville, NY 14221
Telephone: (585) 626-9303
Web site: www.mafell.com
E-mail: mafell@msn.com
Contact: Dennis Hambruch

For more than seventy years, Mafell has stood for reliable and high-quality portable woodworking machinery, including woodworking machines from Germany.

Timber Tools
304 Carlingview Drive, Toronto, ON M9W 5G2
Telephone: (416) 675-2366
Web site: www.timbertools.com
E-mail: Gary Richter at gary@niagaratech.com

Professional power tools for cutting, mortising, and slotting.

Timberwolf Tools
P.O. Box 258, Freeport, ME 04032
Telephone: (207) 865-7095 or (800) 869-4169
Web site: www.timberwolftools.com
E-mail: info@timberwofltools.com
Contact: Darren Bantz

Offers a full line of specialty power tools from Makita, Oliver Machine Company, Hitachi, and German Engineered HOLZ-HER System tools by Protool.

schools

Black Rapids Timber Framing School
1307 Windfall Way, Fairbanks, AK 99709
Telephone: (907) 455-6158
Web site: www.blackrapids.org
E-mail: startips@ptialaska.net
Contacts: Annie and Michael Hopper

College of the Rockies
555 McKenzie Street, Kimberley, BC V1A 2C1
Telephone: (250) 427-7116
Web site: www.cotr.bc.ca/kimberley
E-mail: kimberley@cotr.bc.ca
Contact: Mike Flowers

Weekend workshops to two-month programs.